Emerging Techniques for Food Processing and Preservation

The demand for safe and healthy foods by consumers has increased interest in developing new food processing techniques over the past decades. Emerging technologies and techniques are not just working to increase the shelf life of food but are also functioning to maintain the same quality of food that makes it desirable in the first place.

Emerging Techniques for Food Processing and Preservation is an essential guide for professionals and researchers in the food industry who seek to stay updated on the latest advancements in food processing and preservation techniques. This comprehensive book explores cutting-edge technologies that can enhance the quality and safety of food products while also improving their shelf life.

With contributions from leading experts in the field, this book covers a wide range of topics, including electrodialysis, refractance window technology, cold plasma, biospeckle laser techniques, nanofluids, and many others. Each chapter includes detailed explanations of the principles behind these emerging techniques, as well as case studies that demonstrate their practical applications.

In this book, readers will gain insights into the principles behind these emerging techniques, their advantages and limitations, and the practical applications in various food products. Whether you are a food scientist, engineer, or a food industry professional, this book will help you stay at the forefront of the rapidly evolving landscape of food processing and preservation.

Emerging Techniques for Food Processing and Preservation

Edited by
Swati Kapoor, Gurkirat Kaur,
B. N. Dar, and Savita Sharma

CRC Press
Taylor & Francis Group
Boca Raton London New York

CRC Press is an imprint of the
Taylor & Francis Group, an **informa** business

Cover image: Shutterstock

First edition published 2024
by CRC Press
2385 NW Executive Center Drive, Suite 320, Boca Raton FL 33431

and by CRC Press
4 Park Square, Milton Park, Abingdon, Oxon, OX14 4RN

CRC Press is an imprint of Taylor & Francis Group, LLC

ISBN: 978-1-032-06847-3 (hbk)
ISBN: 978-1-032-10804-9 (pbk)
ISBN: 978-1-003-21713-8 (ebk)

DOI: 10.1201/9781003217138

Typeset in Century Old Style Std
by Apex CoVantage, LLC

Contents

CONTENTS

Preface

Emerging technologies and techniques are not working only to increase the shelf life of food; they are also functioning to maintain the same quality of food that makes it desirable in the first place. The demand for safe and healthy foods by consumers has increased interest in developing new food processing techniques over the past decades.

This book aims to provide a comprehensive overview of the latest advancements in the field of food science and technology. The book focuses on innovative techniques such as electrodialysis, refractance window technology, ozone technology, and more and explores their principles, advantages, and limitations.

The book features expert contributions from leading researchers and practitioners in the field, providing readers with accurate and up-to-date information. Each chapter covers a specific emerging technique, providing in-depth analysis and practical examples of its application in various food products.

These topics are increasingly important in the field of food science and technology and provide readers with a comprehensive understanding of the latest developments. The book's clear and concise presentation of complex concepts makes it accessible to a wide range of readers, including food scientists, researchers, and students.

We hope that this book will serve as an essential resource for those interested in staying up to date with the latest advancements in the field of food processing and preservation. We also hope that it will inspire further research and innovation in this field, leading to the development of new and improved techniques that will benefit the food industry and consumers worldwide.

About the Editors

Dr. Swati Kapoor, M.Sc., PhD, is a food technologist at Punjab Horticultural Postharvest Technology Centre, Punjab Agricultural University, Ludhiana, Punjab, India. Currently, the scientist is involved in developing sustainable technologies from fruit and vegetable wastes; management of quality control laboratory for fresh and processed foods; and shelf-life extension of fresh fruits and vegetables using edible coatings, low temperature, packaging, and so on.

Dr. Gurkirat Kaur, M.Sc., PhD, is a scientist (Nanotechnology) working in the Electron Microscopy and Nanoscience Laboratory at Punjab Agricultural University. Dr. Kaur has experience in teaching and research in food science and technology for the last ten years. Dr. Kaur's responsibilities include scientific and technical leadership and development of advanced facile technologies for food processing and protection.

Dr. Basharat Nabi Dar, PhD, is an assistant professor in the Department of Food Technology at the Islamic University of Science & Technology, Awantipora, JK, India. He is the recipient of the prestigious CV Raman Fellowship and UGC Research Award in the field of agricultural sciences by the University Grants Commission, New Delhi, GOI. He is a graduate of agricultural sciences. He has published more than 100 research and review papers in his field in various prestigious national and international journals and presented papers at many conferences. The main focus of his research is on agricultural by-products, traditional foods, and unexplored food resources.

Prof. Savita Sharma is a principal scientist (dough rheologist) working in the Department of Food Science and Technology at Punjab Agricultural University. She has been actively involved in teaching and research for more than 35 years. Her research mainly focuses on the potential utilization and value addition of cereal products; functionality enhancement of pasta; and processing and utilization of cereal brans, non-conventional legumes, and novel grains. She has guided 30 M.Sc. students and 12 PhD research scholars. Prof. Sharma has published more than 100 research papers in reputed national and international journals and also edited and authored books in food science and technology.

Contributors

Dolly Bhati
TEAGASC Food Research Center, Ashtown
Dublin, Ireland

Reshma Bhatnagar
Sarbi Petroleum & Chemicals Pvt Ltd, Navi Mumbai
Maharashtra, India

B. N. Dar
Islamic University of Science and Technology, Awantipora
Jammu and Kashmir, India

Isha Dudeja
Punjab Agricultural University, Ludhiana
Punjab, India

Srinivas Girijal
University of Horticultural Sciences, Bagalkot
Karnataka, India

Antima Gupta
Punjab Agricultural University, Ludhiana
Punjab, India

Muskaan Gupta
Punjab Agricultural University, Ludhiana
Punjab, India

Kirti Jalgaonkar
ICAR-Central Institute for Research on Cotton Technology, Mumbai
Maharashtra, India

Aakriti Kapoor
Punjab Agricultural University, Ludhiana
Punjab, India

Swati Kapoor
Punjab Agricultural University, Ludhiana
Punjab, India

Gurkirat Kaur
Punjab Agricultural University,
   Ludhiana
Punjab, India

Randeep Kaur
Punjab Agricultural University,
   Ludhiana
Punjab, India

Manoj Kumar Mahawar
ICAR-Central Institute for Research
   on Cotton Technology, Mumbai
Maharashtra, India

Darakshan Majid
Islamic University of Science and
   Technology, Awantipora
Jammu and Kashmir, India

H. A. Makroo
Islamic University of Science and
   Technology, Awantipora
Jammu and Kashmir, India

Mehrajfatema Zafar Mulla
Kuwait Institute for Scientific
   Research
Safat, Kuwait

L. Muthulakshmi
Kalsalingam Academy of Research
   and Education, Krishnankoil
Tamil Nadu, India

H. R. Naik
Sher-e-Kashmir University of
   Agricultural Sciences and
   Technology,
Jammu and Kashmir, India

Saadiya Naqash
Islamic University of Science and
   Technology, Awantipora
Jammu and Kashmir, India

Pooja Nikhanj
Punjab Agricultural University,
   Ludhiana
Punjab, India

Priyanka
Punjab Agricultural University,
   Ludhiana
Punjab, India

R. Rajam
Kalsalingam Academy of Research
   and Education, Krishnankoil,
Tamil Nadu, India

Sudha Rana
Punjab Agricultural University,
   Ludhiana
Punjab, India

S. Mohan
Kalsalingam Academy of
   Research and Education,
   Krishnankoil
Tamil Nadu, India

Arashdeep Singh
Punjab Agricultural University,
   Ludhiana
Punjab, India

Prastuty Singh
Punjab Agricultural University,
   Ludhiana
Punjab, India

Rajan Sharma
Punjab Agricultural University,
   Ludhiana
Punjab, India

Savita Sharma
Punjab Agricultural University,
   Ludhiana
Punjab, India

Chapter 1

Introduction to Emerging Technologies in the Food Processing Sector

Swati Kapoor, Muskaan Gupta, Sudha Rana, Gurkirat Kaur, B. N. Dar, and Savita Sharma

1.1 Introduction

Consumers today seek high-quality foods that have minimal nutrient loss, are microbiologically safe, and are ready to eat. Consumers are becoming more aware of the health advantages and risks linked with food product *viz.* microbial, physical, and chemical concerns. New technologies are being created to fulfil consumer expectations. While thermal processing ensures food safety, it also causes the destruction of certain food characteristics as well as nutrient degradation. Food texture, flavour, aroma, colour, and nutritive values are all important factors in a consumer's decision to purchase and accept a product. Colour, texture, aroma, and flavour are irreversibly affected, necessitating the creation of new technologies. As a result, in most food processing technologies, overall food quality enhancement like safety, security, nutritional value, and processing cost reduction have always been a focus.

DOI: 10.1201/9781003217138-1

1

Emerging technologies in food processing are the focus of development to meet specific consumer demands for healthful, safe, and minimally processed foods. Several alternative processing and transport methods have been created in recent years; however, some are yet to be industrialized, or their optimization is still dependent on empirical methodologies. These revolutionary procedures also contribute to low-energy food manufacturing approaches that are environmentally sustainable. Also, these practices play a role in diminishing water consumption, which circumvents some of the constraints inflicted by modern food processing principles (Toepfl et al., 2006). These are the primary motivators for food science and technology to create alternative methods to improve food safety. As a result, it is clear that food sector innovation is a societal challenge with great implications for long-term competitiveness, creation, and advancement.

Making the most of the explicit possibilities of these new technologies, including study and control of complicated process-structure-function interactions, allows for the science-based development of custom-made foods.

1.2 Role of Advanced Technologies in Food Processing

Food processing companies are working harder than ever to improve food quality and safety around the world. The new consumer is facing different demands from the past like healthy, convenient, safe, fresh-quality, and sustainable foods. There has been a significant increase in the number of outbreaks of food-borne disease in recent years, which has been a serious public health problem. As a result, both the food industry and consumers are concerned about microbiological food safety. Suitable alternatives to conventional methods are helping throughout the processing, distribution, and supply chain to maintain the quality indices of the product in terms of sensory attributes and nutritional characteristics, along with prevention of microbial and fungal deterioration. Forward osmosis, MRI, ozone, nanobiosensors, and bio-based composites are some of the advanced technologies that can be utilized to epitomize scalable and flexible food manufacturing processes.

1.2.1 Forward Osmosis

Forward osmosis (FO) is also referred to as "direct osmosis", "engineered osmosis", or "manipulated osmosis". FO drives water permeation from a lower-concentration solution (feed solution) into a concentrated solution (draw solution) using osmotic pressure generated by various solute

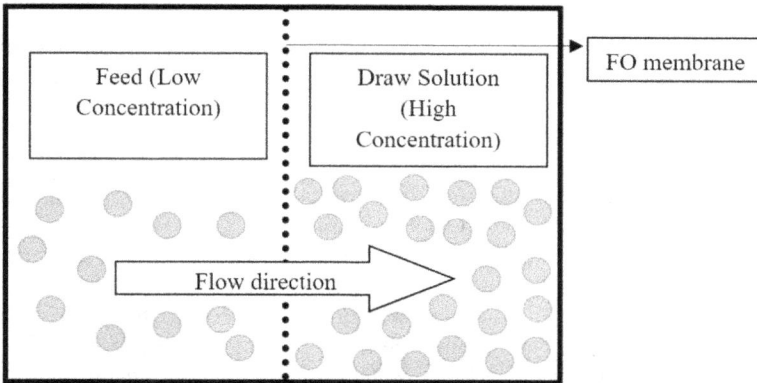

Figure 1.1 Mechanism of forward osmosis.

Source: Adapted and redrawn from Rastogi (2016)

concentrations using a membrane (constructed of polymeric materials) that acts as a barrier, allowing water to pass through while obstructing the molecules that are big, including suspended solids like dissolved molecules or ions. As a "soft" concentration process, forward osmosis has gained appeal. The technique has observed a growing interest in past years due to basic equipment and simple configuration with low energy consumption. The enhanced performance and preservation of nutritional compounds due to mild operating conditions acted as the stimulus behind the growing interest. Besides the ability to operate without a significant pressure, it has high water recovery and able to feed with high solid content with less fouling (Xu *et al.*, 2010; Rastogi, 2016; Sant'Anna *et al.*, 2012). The forward osmosis technique has been widely used in seawater desalination, the food processing industry, the pharmaceutical industry, and wastewater treatment. As there is a great demand for a concentration technique that can preserve bioactive-containing fluids, the technique is generally utilized in the food processing industry for concentration of beverages along with retention of nutritional and sensorial attributes.

1.2.2 Magnetic Resonance Imaging

MRI is a non-destructive, non-ionizing technology employing atomic particles that interact with an external source of a magnetic field to emit energy at a certain frequency. It is mostly employed in diagnostic radiology. The imaged tissue structures are indicated by the emission signals. In the same way, MRI can be used in food science to image the internal structures of food. The techniques of magnetic resonance imaging are not only extraordinary because of their eloquent and reliable fundamentals but also because of their online identification and authentication of food processing characteristics, providing information into the underlying real-time mechanisms and dynamics of the process.

Due to many technological improvements and developments, MRI, also known as an image-based approach, has drawn widespread interest in the detection of food quality and safety and has become a vital instrument in all areas of application. As an imaging technology, MRI uses a magnetic field and a radio frequency system to convey the signal characteristics of a product, allowing physical and chemical information to be seen. When a food product containing a lot of water is exposed to a strong magnetic field, the average magnetic moment of many protons aligns with the field's direction, and magnetic characteristics are developed. Three-dimensional analysis can be performed on MRI images since they have a higher contrast than standard camera images. The cost of an MRI is high, and the analysis is based on protons (1H nuclei) or a body that is only made up of water.

1.2.3 Ozone

The allotropic form of oxygen that is created from oxygen during lightning or UV irradiation events is known as ozone and is a powerful oxidizing agent in nature. O_2 splits into extremely reactive singlet oxygen during ozone formation, which then interacts with other oxygen molecules to generate ozone. The ozone molecule is made up of three oxygen atoms, and its unpaired electrons are arranged around an oxygen nucleus in the core, giving it high reactivity. Under proper conditions, ozone is a safe disinfectant. Ozone is suitable in food processing and preservation, as ozone decomposes quickly to O_2 with few residual effects that are short lived. Some combinations of ozone with other technologies, like pasteurization, ultraviolet radiation, high-pressure processing, and membrane processing, may be remarkably useful in enhancement of shelf life and ensuring microbial safety.

1.2.4 Nanotechnology

Nowadays, nanotechnology is one of the major emerging technologies that is setting foot out of the laboratory into every sector of food production. Commercial applications of nanotechnologies are widely used in the food industry, packaging, and storage. Commercially, nanofoods are estimated to vary widely between 150–600 nanofoods and 400–500 nanofood packaging applications. There are numeral nanomaterials used for different food applications on an industrial scale. Applications of nanotechnology in food industries include nanoparticulate delivery systems (nanodispersions and nanocapsules), packaging (nanolaminates, nanocomposite bottles, bins with silver nanoparticles), food safety and biosecurity (nanosensors), and others (Chen *et al.*, 2006). Nanoscale food additives can also be used to alter nutritious improvement, texture, and flavour; improve functional characteristics; and even detect pathogens. In food packaging, application of nanotechnology includes edible, nano wrapper which are used to envelope foods, prevent gas and moisture exchange and develop other "smart" packaging (containing nanosensors and antimicrobial activators) for detection of food spoilage and release of nano-anti-microbes to extend food shelf life (Nickols-Richardson and Piehowski, 2008; Miller *et al.*, 2008).

1.2.5 Nanobiosensors

Food preservation, in addition to food packaging, is extremely important in the food industry. Food spoilage can be detected using nanobiosensors, which are made up of hundreds of nanoparticles that fluoresce in different colours when they come into contact with pathogens. They are an electronic data processing device with a sensing component that can detect and convert changes in heat, humidity, light, gas, and chemicals into electrical signals. Nanosensors are more efficient than conventional sensors due to their great sensitivity and selectivity.

Besides food packaging, food preservation is also of substantial importance for the food industry. The main aim of nanosensors is reduction of time for pathogen detection from days to hours or even minutes. Nanosensors could be directly placed into packaging material, where they would serve as "electronic noses" or "tongues" by detecting chemicals released during food spoilage. Numerous types of nanosensors are used, such as array biosensors, nanotest strips, nanocantilevers, nanoparticles in solution, nanoparticle-based sensors, and electronic noses.

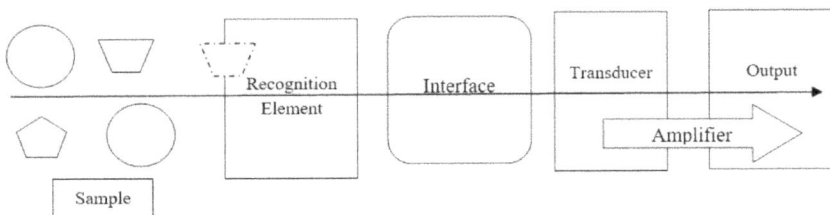

Figure 1.2 Working principles of biosensors.

- A few nanosensors are under development; for example, modified carbon nanotubes as a biosensor to detect microorganisms, toxic substances, and spoilage of foods or beverages are being developed at Georgia Tech in the United States.
- Opel was created by Opalfilm with 50-nm carbon black nanoparticles and was also employed as a biosensor that changes colour in reaction to food deterioration.
- A digital transform spectrometer (DTS) was produced by Polychromix (Wilmington, MA, USA) that uses microelectromechanical system technology to detect trans fat content in foods.

Nanocantilevers are another innovative class of biosensor. Their principle of detection is based on their ability to detect biological binding interactions, such as interaction between enzyme and substrate, antigen and antibody, cofactor and receptor and ligand, through physical and/or electromechanical signalling. The assembly consists of tiny pieces of silicon-based materials that have the capability to recognize proteins and detect pathogenic bacteria and viruses. Enormous success of nanocantilever devices has been observed in studies of molecular interactions regarding detection of contaminant chemicals, toxins, and antibiotic residues in food products. The silicon surface of nanocantilevers can be modified to attach antibodies, resulting in a change of the resonant frequency depending on the attached mass. With the utilization of nanogas sensors, nanosmart dust can be used to detect any sort of environmental pollution. These sensors are composed of small wireless sensors and transponders. Nanobarcodes are also considered to be an efficient mechanism for the detection of the quality of agricultural produce.

Biomimetic sensors and smart biosensors are developed using protein and biomimetic membranes which have been reported to efficiently determine the presence of mycotoxins and other toxic compounds. In packaging, material made of biodegradable nanocomposites has a lot of environmental promise. Smart packaging helps customers select products with a long shelf life while also indicating the food's nature and other qualities. Food safety and security are also enhanced by this technology. The incorporation of nanocomposites, nanosensors, and biodegradable nanocomposites in food packaging has transformed packaging by making it leak proof, gas free, and pathogen free. They serve as a barrier for gas exchange and protect food quality using nanoclays, while antimicrobial packaging employing silver, titanium oxide, zinc oxide, and other bionanoparticles prevents the formation of microorganisms, as well as pathogens and toxins.

Montmorillonite is a reasonably inexpensive and commonly available natural clay formed from volcanic ash/rocks that is utilized in nanocomposites for food packaging. Nanoclay has a natural nanolayer structure that inhibits gas permeability and improves the gas barrier characteristics of nanocomposites significantly. The polymers used for clay–polymer nano-composites are polyamides (PSs), polyolefins, nylons, polystyrene (PS), ethylene-vinylacetate (EVA) copolymer, epoxy resins, polyimides, polyure-thane, and polyethylene terephthalate (PET).

Guard IN Fresh is another commercialized nanocomposite-based product that aids in the ripening of fruits and vegetables by scavenging ethylene gas (Gupta and Moulik, 2008). The Top Screen DS13 nanocomposite is an example of an easily recyclable nanocomposite. Top Screen DS13, unlike wax-based coatings, promotes the idea of being water based and thereby quickly destroyed (Mason *et al.*, 2006).

1.2.6 Electrodialysis

Electrodialysis is an electro membrane method which is based on the principle of dialysis that modifies the ionic component of fluids using ion-permeable (selective) membranes under the action of a constant electrical potential. In electrodialysis, the membranes used either have a constant positive charge or a constant negative charge or are referred to as anion or cation exchange membranes, respectively. When an explicit current is applied at the electrodes in electrodialysis, the electrical charges on ions enable them to be directed through solutions of membranes, with positively charged cations migrating toward the negatively charged cathode and negatively charged anions migrating toward the positively charged anode. Cations will pass through cation-exchange membranes under the influence of an electrical potential, while negatively charged anions will pass through anion exchange membranes but be trapped by negatively charged cation-exchange membranes. Electrodialysis has various applications in food processing, for example, demineralization of food products (whey, removal of ions from syrups, juices) and wine stabilization.

1.2.7 Refractive Window Technology

Lately, refractive window (RW) drying is a cutting-edge drying technology and has been used for the drying of food products where the temperature of the food product is kept below 70°C. This technology dehydrates moist food products upon a thin Mylar sheet that is transparent to infrared light and serves as a drying "window". It improves the drying process, in which hot water transfers heat energy to a moist substance by causing moisture to evaporate. Water is used as the heat source in the refractive window dryer, and all heat transfer modes, including conduction, convection, and radiation, are operational. Conduction is the most common method of heat transfer, while the other two modes are determined by the resistance of water available in each mode. Despite radiation being an innate and rapid process, major heat loss from the water to the outside occurs by conduction. The circulating water temperature in the RW dryer is kept between 95 and 98°C. During drying, a Mylar film is placed over the hot water surface to prevent water evaporation and heat loss. As a result, the temperature of the product remains at 60–72°C.

1.2.8 Biomimicry

Biomimicry is the imitation or copying of an environment in all of its aspects, methods, and operations in order to solve the world's most important problems. Biomimicry strategies have been shown to improve sustainability and efficiency, particularly in the domains of food and agriculture. The physico-chemical features of the processed food, as well as the role of lipids in the food structure, limit the capacity to eliminate trans fats without adding extra saturated fats. Alternative techniques for maintaining levels of cardio-protective unsaturated fat are needed. The use of oleogels, or small-molecule molecular gels, is one such technique (Wang and Rogers, 2015).

1.2.9 Cold Plasma

Non-thermal plasmas are characterized by electron temperature much higher than that of heavy particles. The electron density of non-thermal plasma is low ($<10^{19}$ m^{-3}). Plasma chemistry has been induced due to inelastic collisions between electrons and heavy particles. Few elastic collisions result in the slightly heating of heavy particles and higher electron temperature (Tendero et al., 2006; Moreau et al., 2008). Plasmas are produced by applying energy to a known volume of neutral gas, resulting in ionization and formation of the fraction of free electrons and ions. In particular, non-thermal plasma can be generated by a number of electrical discharges or electron beams, and input energy is provided in the form of electrical energy, whereby the electrical energy is first transformed into energetic electrons rather than heating the whole gas. This leads to the formation of excited species, free radicals, and ions through the electron-impact dissociation, excitation, and ionization of the background gas molecules (Fridman et al., 2005; Foest et al., 2006).

Formation of cold plasma using humid air leads to the production of active species such as OH- and NO- radicals (Benstaali et al., 1998; Laroussi et al., 2003). These radicals cause intense bombardment around microorganisms, causing surface lesions that cannot be repaired and ultimately lead to the destruction of cells. This whole process is known as etching (Pelletier, 1992). Moreover, UV photons, the active reactive species of plasma, can alter the DNA of micro-organisms and lead to disrupted cell replication. In addition to reactive species, UV photons can modify the DNA of microorganisms and as a result disturb cell replication (Boudam *et al.*, 2006).

1.2.10 Biospeckles

Biospeckles are a non-destructive optical tool used to examine biological materials. The principle behind the technology is generation of a speckle pattern is generated in an observation plane due to the interference of back-reflected light. The speckle pattern so formed, generally has two modules: a static pattern and variable pattern. Static pattern is generated by the tissue's fixed particles and a variable one is generated by the tissue's moving sections. The term "biospeckle" is used for the changeable speckle pattern that is a distinguishing property of biological material (Pandiselvam *et al.*, 2020). This approach has recently seen widespread use in the agriculture sector for evaluating the quality and safety of food products due to its potential for use in the food business to determine quality of sample without even have to touch them along its simplicity in use and low cost.

1.2.11 DNA Foil Technology

DNA foil is a portable, entirely self-administered, on-site DNA test that confirms food adulteration detection in as little as 30 minutes without the use of expensive polymerase chain reaction (PCR) setup. This test basically consists of five different steps. In first and second stages, it contains a barrel that breaks down, lyses, removes, neutralizes, and stabilizes DNA from a variety of food samples. As a result, the preparation and DNA extraction steps from the sample can be completed with only a few actions. In step 3, that is amplification stage for the DNA target, is carried out by transferring one drop of extracted DNA to the reaction tube and placing it in water. Next, the DNA targets are amplified and hence, multiple copies are produced by employing enzymes and specific primers are identified. After 30 minutes, the final step will be the detection step in which a revealing strip has to be inserted, which allows transport of target DNA (El Sheikha, 2019).

1.2.12 E-Sensing

E-noses, e-tongues, and e-eyes are examples of human-based sensing devices that can be used to examine different substances in a variety of food matrices. Electronic sensing (e-sensing) is a phrase that describes devices that are based on three of the five senses of the human body. An e-nose is a sensor that can distinguish between human sensory qualities like aromas and flavours. The unit is composed of a sensor system that detects gases and odours using multiple chemosensors. An e-eye is derived from a computer image analysis technology known as computer vision, which converts photos into digital photos or obtains them from colour sensors that replicate human vision and are used to distinguish and classify objects. An e-tongue detects tastes based on taste bud sensations, which are linked to membrane receptors on the tongue that identify flavours in various different spots (Galvan *et al.*, 2022).

1.3 Effects of Advancements on Food Safety and Quality Assurance

Food is a vital determinant of health, nutritional status, and productivity of the population. Therefore, it is essential that the food we consume be safe and wholesome. Unsafe food can lead to a various food-borne diseases. Food safety and quality control are not only important at the household level but are critical in large-scale food production and processing and also where fresh food is prepared and served. Food safety is a concept/method/procedure that food will not cause any harm when it is being treated or eaten. There are number of food practices such as Good manufacturing practices (GMP), Hazard analysis critical control points (HACCP), and International organization for Standardization (ISO), and a number of food industries have adopted different types of food practices in order to ensure that their foods are safe for consumption. Safe food is a result of food safety practices or an product of food safety techniques. A food which is free of hazards, including chemical, physical, and biological hazards, has no potential to cause adverse effects.

Advancements in food processing technologies, larger per capita incomes, increased purchasing power, and also consumer demand, which has increased for a variety of foods like extruded products, functional foods, and health foods, have increased the manufacturing of these foods. The safety of such foods has to be assessed. The quality of foodstuffs, both raw and processed, is of public health concern and should be addressed. The safety of processed and packaged foods should be addressed. Various applications of emerging technologies in the food sector are listed in Table 1.1.

Traditionally, fruit juice is extracted employing thermal methods. However, it is difficult to maintain the food nutrients, such as vitamin C, in such operations. As most nutrients are sensitive to heat, exposing them to high temperatures during concentration of juices can cause degradation in bioactive components. Among the alternatives, FO has a number of appealing properties that exceed the disadvantages of thermal-based procedures.

Table 1.1 Application of Emerging Advanced Technologies in the Food Processing Industry

Advanced Technology	Product	Application	References
Forward osmosis	Blueberry juice	Concentration of juice from 15°B to 17–18°B Uses citric acid, potassium sorbate and sodium benzoate as draw solution	Chu *et al.*, 2022
	Microalgae	Dewatering with average rate 2 l/m² membrane/hr with decrease in volume up to 65–85%	Buckwalter *et al.*, 2013
	Grapefruit juice	Concentration of juice from 11.4°B to 12.4, 16.5, 22.2, and 45.1°B at 10, 30, 50, and 75% concentrated juice, respectively Using NaCl, glucose as draw solution and pressure-assisted draw solution (at 50% concentration	Kim *et al.*, 2019
	Betacyanin from *Opuntia ficus-indica*	Concentration of bioactive compound from 898 to 1325 mg/l Maximum concentration using NaCl and 50°C temperature	Ravichandran and Ekambaram, 2018
	Skim milk and whey	Concentration using NaCl as draw solution (48–57 g/l) Resultant total solids achieved 21 and 15% in skim milk and whey, respectively	Chen *et al.*, 2019
	Anthocyanin extract from *Garcinia indica*	Concentration of anthocyanin from 49.68 to 2692 mg/L with lower non-enzymatic browning and lower rate of conversion from HCA to its lactone form as compared to thermal concentrate	Nayak and Rastogi, 2010

(Continued)

Table 1.1 (Continued)

Advanced Technology	Product	Application	References
Magnetic resonance imaging	Fish (fresh and frozen thawed cutlets)	Quantification of muscle fat content and subcutaneous adipose fat	Collewet et al., 2013
	Pig carcasses and cuts	Quantification of muscle, subcutaneous fat, and muscular fat using low-field MRI	Monziols et al., 2006
	Atlantic salmon (prerigor, postrigor, frozen, thawed)	Assessment of water mobility and salt uptake in relation to microstructure of fish muscle ^1H and ^{23}Na MRI	Aursand et al., 2009
	Raw potatoes	Assessment of imaging features for prediction of textural sensory attributes	Thybo et al., 2004
	Tomato fruit	Monitoring ripening of tomato during post-harvest storage of 3 weeks by detecting air space and gas bubble content	Musse et al., 2009
Ozone	Whole powder and skim milk powder	Inactivation of Cronobacter	Torlak and Sert, 2013
	Wine grapes	Postharvest fumigation before dehydration to reduce yeast and fungi	Botondi et al., 2015
	Apple juice	Inactivation of E. coli 0.157:H7, Salmonella typhimurium, Listeria monocytogenes	Choi et al., 2012
	Stored wheat	Reduction of 96.9% fungal spores by 5 min of ozonation	Wu et al., 2006

Technology	Sample	Description	Reference
Nanobiosensors	Soft wheat grains	Reduction of total fungal content and mycotoxins (deoxynivalenol and total aflatoxins)	Trombete et al., 2017
	Cacao pods	Detection of Phytophthora palmivora causing black pod rot	Franco et al., 2019
	Honey and milk	Detection of sulfadiazine	He et al., 2020
	Chicken	Detection of Salmonella typhimurium using optical-based biosensor	Zheng et al., 2020
	Milk	Detection of melamine using fluorescent copper nanomaterial	Ga et al., 2020
Electrodialysis	Whey	Demineralization	Houldsworth, 1980
	Synthetic UF-whey permeate and UF-whey retentates	Demineralization	Pérez et al., 1994
	Wines	Stabilization of wine by removal of potassium hydrogen tartrate	Gonçalves et al., 2003
	Passionfruit juice	Deacidification up to pH 4.5	Vera et al., 2009
	Cranberry juice	Deacidification	Serre et al., 2016
Refractive window technology	Milk	Concentration of milk up to 26% at 0.45 bar and 70°C	Al-Hilphy et al., 2022
	Peach palm-tucupi blend	Dehydration	da Costa et al., 2019
	Sapota bar	Drying of sapota pulp for preparation of bar	Jalgaonkar et al., 2020

(Continued)

Table 1.1 (Continued)

Advanced Technology	Product	Application	References
	Unripe green banana flour	Drying through refractive window drying and hot air drying Retention of better quality in refractive window drying than hot air drying	Padhi and Dwivedi, 2022
	Paprika cv. jalapeno	Drying and assessment of carotenoids and Scoville heat units in comparison to freeze drying, oven drying, and natural convective drying	Topuz et al., 2011
Cold plasma	Duck eggs	Bactericidal effects of cold plasma against Salmonella enterica treated by arc plasma for 10–40 seconds	Gavahian et al., 2019
	Rapeseeds	Reduced contamination (moulds, yeast, Bacillus cereus, E. coli, Salmonella spp.) Increased germination rate and seedling growth	Puligundla et al., 2017
	Siriguela juice	Improved content of pigments, total phenolics, antioxidant activity, and vitamin B contents	Paixão et al., 2019
	Food packaging material	Surface disinfection of glass, polyethylene, polypropylene, nylon, and paper foil Reduction in E. coli, Salmonella typhimurium and Staphylococcus aureus	Puligundla et al., 2016
	–	Inactivation of polyphenol oxidase and peroxidase enzyme	Surowsky et al., 2013
Biospeckle	Beef	Assessment of biological activity due cathepsins and calpains in sample of muscle during aging	Amaral et al., 2013

	Sample	Objective	Reference
	Kefir grains	To analyse viability of kefir grains and their state during beverage production	Guedes et al., 2014
	Mango cv. Tommy Atkins	Assessment of mechanical damage during maturation for 21 days	Santana et al., 2021
	Beans	Detection of fungi in seeds	Braga Jr et al., 2005
	Apples cv. red delicious	Assessment of mealiness	Arefi et al., 2017
	Maize	Assessment of presence of vitreous and floury endosperms	Weber et al., 2014
E-sensing	Hops	Assessment of aging of pellets and discriminate between different types using e-nose	Lamagna et al., 2004
	Bread	Detection of mould using E-nose and KNN classifier	Estakhroueiyeh and Rashedi, 2015
	Fruit juices	To detect adulteration / To classify pure juice and industrial juice samples using e-nose	Rasekh and Karami, 2021
	Coffee	To classify roasted coffee on the basis of acidity levels using e-nose	Thazin et al., 2018
	Sapodilla tea	Determination of sensory attributes using e-tongue	Rahimah et al., 2020
	Bovine and goat milk	Detection of change in taste using array of lipid/polymer-based membrane-based e-tongue	Tazi et al., 2017

Owing to the benefit of using low temperatures, forward osmosis can be employed to maintain the sensory and nutritional qualities of fruit juices at low temperatures and pressures. Because of its low-temperature operation, FO can create high-concentration fruit juice and retain its nutritive content. Milczarek *et al.* (2020) showed improved quality of watermelon juice upon concentration up to 65°B (Brix) by forward osmosis in comparison to conventional methods. L^*, a^*, and b^* values elucidated less darkening in FO concentrate than thermal concentrate. Also, antioxidant activity in reconstituted FO concentrate was significantly higher than in reconstituted thermal concentrate. However, no significant differences were reported in citrulline content, lycopene content, and total soluble phenolics. Aroma profiling of FO concentrate was comparable to "fresh-cut watermelons", in contrast to thermal concentrate, which was associated with "fishy" and "mushroom" off-odours (Milczarek and Sedej, 2021). Nayak and Rastogi (2010) also elucidated that non-enzymatic browning (NEB), ratio of HCA lactone to lactone, and the degradation constant were much higher in thermally concentrated anthocyanin extract in comparison to forward osmosis–concentrated anthocyanin extract from *kokum* (*Garcinia indica*).

Internal food tissues, defects, and quality can be visualized using MRI. Over or under processing of food products can alter the final result, according to MRI research. Furthermore, sample delays and traditional methods of product analysis during the process lengthen the process time and may result in inaccurate results. Because of its non-destructive properties and capacity to offer highly detailed spatial information on the distribution and microenvironment of water in soft tissues, magnetic resonance imaging is effective for managing fruits (Hancock *et al.*, 2008). Pre- and post-harvest research on evaluating physical and biological changes in fruits could benefit from the use of magnetic resonance imaging devices. Online identification of internal defects and water core detection are examples of sensor-based applications. Herremans *et al.* (2014) described that the water core in apples can be successfully detected through MRI by detecting tissue with higher water content in different apple cultivars with superior images in comparison to X-ray CT. As a non-invasive technique, MRI based sensors could be used for real-time monitoring of water spatial dynamics in processes such as drying, freezing, cooking, and freeze-drying (Kirtil *et al.*, 2017). Song *et al.* (2017) detected water dynamics in abalone during drying and rehydration process. Similarly, Song *et al.* (1992) assessed moisture transfer through maize kernels during drying and established that there are two primary routes of moisture transfer through maize kernels: through pericarp and through the glandular layer of scutellum. Evans *et al.* (2002) also studied spatial changes in water mobility, water content, and structural changes in strawberry during osmotic dehydration and air drying.

Spoilage in food can be well detected with use of nanosensors. For example, food pathogens can be detected by employing an array of thousands of nanoparticles designed to fluoresce in different colours on contact with them. These type of sensors are usually sensitive towards gases such as hydrogen, hydrogen sulphide, nitrogen oxides, sulphur dioxide, and ammonia. It is a device comprising an electronic data processing part and a sensing part that is able to convert any change in light, heat, humidity, gas, and chemicals into electrical signals. These nanosensors are considered to be more efficient than conventional sensors due to its high sensitivity and selectivity. Gas sensors are generally made up of metals such as palladium, platinum, and gold. Gold-based nanoparticles are also used at times to detect toxins such as aflatoxin B1 in milk. Kumar *et al.* (2017) listed numerous studies that employed nanosensors for detection of food pathogenic bacteria, food-contaminating toxins, food-contaminating pesticides and chemicals, and food freshness. Nanosensors can also be employed for quality assessment of food during storage and unstable key food ingredients during processing and storage of food.

A time-temperature indicator/integrator (TTI) helps in detection of the spoilage of food based on the history of temperature. Commercially, TTIs are available as iSTrip, manufactured by the company Timestrip, which are able to detect changes in the temperature of frozen foods and indicating it by colour change. During preservation of frozen foods, TTI, made up of gold nanoparticles, changes its colour to red due to sudden surge in temperature during storage, are used. Other sensors have also been developed to detect presence of *E. coli* in packaged foods, based on reflective interferometry. In presence of contamination, protein of *E. coli* placed on the silicon chip binds to similar proteins. It is based on the principle of scattering of light by mitochondria. The scattered light can be detected by analysing digital images.

Antimicrobial packaging employing silver, zinc oxide, titanium oxide, and other bio-nanoparticles prevents the formation of microorganisms, as well as pathogens and toxins.

Integration of nanocomposites, nanosensors, and biodegradable nano composites has revolutionized packaging industry by manufacturing leak proof, gases free, and pathogen free food packaging. They act as barriers for the exchange of gases and are able to maintain the quality of food using nanoclays. Antimicrobial packaging employing silver, zinc oxide, titanium oxide, and other bionanoparticles prevents the formation of microorganisms, as well as pathogens and toxins.

Most recently, electrodialysis has been used for the production of hydrogen from food waste due to a unique combination of anaerobic digestion and electrodialysis. Electrodialysis improved hydrogen output by a factor of 3.5. The kinetics of substrate consumption and volatile fatty acid synthesis were affected by electrodialysis (Hassan *et al.*, 2019). Customers value fruit juices for their aroma and flavour, but the high acidity of some of them prevents them from being used as components in a variety of products such as ice creams, marmalades, and beverages. As a matter of fact, some acidic fruit juices can be deacidified in order to make them more suitable for use in cuisine. As a result, electrodialysis proves a more environmentally friendly method of deacidification of fruit juices than current methods. Electrodialysis using a bipolar membrane could also be utilized to inhibit enzymatic browning in cloudy apple juice by lowering the pH of juice to 2.0. Also, this technique could be employed for the production of soy protein isolates through iso-electric precipitation. Lowering the pH to 4.2–4.6 results in selective separation of proteins (Bazinet *et al.*, 1998).

Refractive window drying is a method of drying in which the product does not come into direct contact with the heat source. When dried in a refractive window dryer, the visual quality of the dried product is maintained better than when dried in a drum drier, spray dryer, or convective dryer. The hue of the freeze-dried product is approximately identical to that of the refractive window–dried product, but freeze-drying is quite expensive and requires skilled labour, and the product formed is very porous. Nutraceutical and bioactive compounds can be dried effectively at a lower temperature in refractive window technology. This perfect technique promises to preserve sensory, nutritional, and bioactive ingredients in heat-sensitive food products by processing at low temperatures and drying quickly. Dried foods with free-flowing character, reduced water activity, and superior storage stability can be achieved with this process. Al-Hilphy *et al.* (2022) concentrated milk through refractive window technology and reported browning index of concentrated milk to be significantly lower than unconcentrated milk, whereas hue and whiteness index were reported to be lower in unconcentrated milk than concentrated milk. Celli *et al.* (2016) have also dried frozen haskap berries via refractive drying technology and reported 92.9% retention of anthocyanins including cyanidin-3-glucoside, cyanidin-3-rutinoside, and peonidin-3-glucoside. Peonidin-3-glucoside and cyanidin-3-glucoside found to have maximum retention in dried product as 100.7% and 94.9%, respectively. However, cyanidin-3-rutinoside found to have least retention (65.1%) upon drying treatment among the mentioned anthocyanins.

Cold plasma has widely been used in the microbial decontamination of foods. Nitrogen and oxygen gas plasma results in the formation of reactive oxygen and nitrogen species (ROS and RNS) such as O., O_2, O_3, OH, NO., and NO_2. These species can disrupt the movement of bio-molecules by acting on the double bond of unsaturated fatty acid in membrane cells. The reactive species act on the polyunsaturated fatty acid (PUFA) that leads to loss a hydrogen atom to form fatty acid radical. Furthermore, the fatty acid radical is oxidized by O_2, which leads to the formation of lipid hydroperoxide. Amino acids and nucleic acids are oxidized with the action of active species, leading to microbial injury or sometimes death (Critzer *et al.*, 2007; Joshi *et al.*, 2011). Oh *et al.* (2017) reported the use of microwave plasma in the inhibition of *Salmonella typhimurium* on radish sprouts. The operating parameters were operating power (900 W), treatment pressure (667 Pa), treatment time (1–20 min), and nitrogen flow rate (1 L/min). The results indicated a 2.6 log_{10} reduction in *Salmonella typhimurium* and 0.8 log_{10} reduction in total mesophilic aerobes. Min *et al.* (2017) reported the inhibition of *E. coli* on bulk romaine lettuce (5 cm dia) using a dielectric barrier discharge (DBD) atmospheric cold plasma (DACP) treatment at 42.6 kV for 22% RH (relative humidity) for 10 min using air. There was a significant decrease in the number of *E.coli* (0.4–0.8 log CFU/g) in the leaf samples in one-, three-, and five-layer configurations without colour change, and insignificant change was found on the surface of leaf samples.

Biospeckle laser technology is a newly developed non-destructive instrument for biological samples. It's quick, simple to use, and cost effective, and it ensures the produce's freshness and safety. The main application is the identification of disease and abnormalities as well as the monitoring of the ageing and maturation process in fresh products. The biochemical changes occurring in the biological material, as well as other intracellular processes, such as organelle movement, cytoplasmic streaming, cell growth and division, and Brownian motility, are determined to be the source of biospeckle action. Researches such as Silva *et al.* (2018) have developed protocols to employ biospeckle laser technology to estimate water activities of food by using pectin and sucrose solutions as models at varied temperatures.

Modern analytical techniques currently provide an appropriate response to worldwide demands for food safety, quality, and traceability, prompting the development of more convincing analytical approaches, such as molecular approaches, for simple and low-cost adulterant identification in food. Although DNA-based techniques have proven the most effective detection instruments for food adulteration, their commercial use still has a long way to go. As a result, there is still a pressing need for unique, quick, easy, powerful, and ubiquitous technology to detect food adulteration. Consequently, DNA foil technology could be a valuable aid in meeting all of these requirements.

E-sensors can detect one or more substance in mixed food samples, and the results can be used for a variety of purposes when various chemometric methods are used. For food quality control, the integration of e-sensing and chemometric technologies is critical. Longobardi *et al.* (2019) characterized grapes based on geographical origin or agronomic practices. The e-nose allowed grapes to be distinguished based on their production method and geographical origin.

References

Al-Hilphy, A. R., Ali, H. I., Al-IEssa, S. A., Gavahian, M., & Mousavi-Khaneghah, A. (2022). Assessing compositional and quality parameters of unconcentrated and refractive window concentrated milk based on color components. *Dairy, 3*(2), 400–412.

Amaral, I. C., Braga Jr, R. A., Ramos, E. M., Ramos, A. L. S., & Roxael, E. A. R. (2013). Application of biospeckle laser technique for determining biological phenomena related to beef aging. *Journal of Food Engineering, 119*(1), 135–139.

Arefi, A., Ahmadi Moghaddam, P., Modarres Motlagh, A., & Hassanpour, A. (2017). Towards real-time speckle image processing for mealiness assessment in apple fruit. *International Journal of Food Properties, 20*(sup3), S3135–S3148.

Aursand, I. G., Veliyulin, E., Böcker, U., Ofstad, R., Rustad, T., & Erikson, U. (2009). Water and salt distribution in Atlantic salmon (Salmo salar) studied by low-field 1H NMR, 1H and 23Na MRI and light microscopy: Effects of raw material quality and brine salting. *Journal of Agricultural and Food Chemistry, 57*(1), 46–54.

Bazinet, L. F. L. L., Lamarche, F., & Ippersiel, D. (1998). Bipolar-membrane electrodialysis: Applications of electrodialysis in the food industry. *Trends in Food Science and Technology, 9*(3), 107–113.

Benstaali, B., Moussa, D., Addou, A., & Brisset, J.-L. (1998). Plasma treatment of aqueous solutes: Some chemical properties of a gliding arc in humid air. *The European Physical Journal Applied Physics, 4*(2), 171–179.

Botondi, R., De Sanctis, F., Moscatelli, N., Vettraino, A. M., Catelli, C., & Mencarelli, F. (2015). Ozone fumigation for safety and quality of wine grapes in postharvest dehydration. *Food Chemistry, 188*, 641–647.

Boudam, M. K., Moisan, M., Saoudi, B., Popovici, C., Gherardi, N., & Massines, F. (2006). Bacterial spore inactivation by atmospheric-pressure plasmas in the presence or absence of UV photons as obtained with the same gas mixture. *Journal of Physics. Part D: Applied Physics, 39*(16), 3494–3507.

Braga Jr, R. A., Rabelo, G. F., Granato, L. R., Santos, E. F., Machado, J. C., Arizaga, R., . . . & Trivi, M. (2005). Detection of fungi in beans by the laser biospeckle technique. *Biosystems Engineering, 91*(4), 465–469.

Buckwalter, P., Embaye, T., Gormly, S., & Trent, J. D. (2013). Dewatering microalgae by forward osmosis. *Desalination, 312*, 19–22.

Celli, G. B., Khattab, R., Ghanem, A., & Brooks, M. S. L. (2016). Refractance Window™ drying of haskap berry–preliminary results on anthocyanin retention and physicochemical properties. *Food Chemistry, 194*, 218–221.

Chen, G. Q., Artemi, A., Lee, J., Gras, S. L., & Kentish, S. E. (2019). A pilot scale study on the concentration of milk and whey by forward osmosis. *Separation and Purification Technology, 215*, 652–659.

Chen, H., Weiss, J., & Shahidi, F. (2006). Nanotechnology in nutraceuticals and functional foods. *Food Technology, 60*(3), 30–36.

Choi, M. R., Liu, Q., Lee, S. Y., Jin, J. H., Ryu, S., & Kang, D. H. (2012). Inactivation of Escherichia coli O157:H7, Salmonella typhimurium and Listeria monocytogenes in apple juice with gaseous ozone. *Food Microbiology, 32*(1), 191–195.

Chu, H., Zhang, Z., Zhong, H., Yang, K., Sun, P., Liao, X., & Cai, M. (2022). Athermal concentration of blueberry juice by forward osmosis: Food additives as draw solution. *Membranes, 12*(8), 808.

Collewet, G., Bugeon, J., Idier, J., Quellec, S., Quittet, B., Cambert, M., & Haffray, P. (2013). Rapid quantification of muscle fat content and subcutaneous adipose tissue in fish using MRI. *Food Chemistry, 138*(2–3), 2008–2015.

Critzer, F. J., Kelly-Wintenberg, K., South, S. L., & Golden, D. A. (2007). Atmospheric plasma inactivation of foodborne pathogens on fresh produce surfaces. *Journal of Food Protection, 70*(10), 2290–2296.

da Costa, R. D. S., da Cruz Rodrigues, A. M., Borges Laurindo, J., & da Silva, L. H. M. (2019). Development of dehydrated products from peach palm–tucupi blends with edible film characteristics using refractive window. *Journal of Food Science and Technology, 56*(2), 560–570.

El Sheikha, A. F. (2019). DNAFoil: Novel technology for the rapid detection of food adulteration. *Trends in Food Science and Technology, 86*, 544–552.

Estakhroueiyeh, H. R., & Rashedi, E. (2015, October). Detecting moldy Bread using an E-nose and the KNN classifier. In *2015 5th International Conference on Computer and Knowledge Engineering (ICCKE)* (pp. 251–255). IEEE.

Evans, S. D., Brambilla, A., Lane, D. M., Torreggiani, D., & Hall, L. D. (2002). Magnetic resonance imaging of strawberry (Fragaria vesca) slices during osmotic dehydration and air drying. *LWT – Food Science and Technology, 35*(2), 177–184.

Foest, R., Schmidt, M., & Becker, K. (2006). Microplasmas, an emerging field of low-temperature plasma science and technology. *International Journal of Mass Spectrometry, 248*(3), 87–102.

Franco, A. J. D., Merca, F. E., Rodriguez, M. S., Balidion, J. F., Migo, V. P., Amalin, D. M., . . . & Fernando, L. M. (2019). DNA-based electrochemical nanobiosensor for the detection of Phytophthora palmivora (Butler) Butler, causing black pod rot in cacao (Theobroma cacao L.) pods. *Physiological and Molecular Plant Pathology, 107*, 14–20.

Fridman, A., Chirokov, A., & Gutsol, A. (2005). Non-thermal atmospheric pressure discharges. *Journal of Physics. Part D: Applied Physics, 38*(2), R1–R24.

Ga, L., Ai, J., & Wang, Y. (2020). AS1411-templated fluorescent Cu nanomaterial's synthesis and its application to detecting melamine. *Journal of Chemistry, 2020*, 1–6.

Galvan, D., Aquino, A., Effting, L., Mantovani, A. C. G., Bona, E., & Conte-Junior, C. A. (2022). E-sensing and nanoscale-sensing devices associated with data processing algorithms applied to food quality control: A systematic review. *Critical Reviews in Food Science and Nutrition, 62*(24), 6605–6645.

Gavahian, M., Peng, H. J., & Chu, Y. H. (2019). Efficacy of cold plasma in producing Salmonella-free duck eggs: Effects on physical characteristics, lipid oxidation, and fatty acid profile. *Journal of Food Science and Technology, 56*(12), 5271–5281.

Gonçalves, F., Fernandes, C., Cameira dos Santos, P. C., & de Pinho, M. N. (2003). Wine tartaric stabilization by electrodialysis and its assessment by the saturation temperature. *Journal of Food Engineering, 59*(2–3), 229–235.

Guedes, J. D. S., Magalhães-Guedes, K. T., Dias, D. R., & Schwan JR, R. B. (2014). Assessment of the biological activity of kefir grains by biospeckle laser technique. *African Journal of Microbiology Research, 8*(27), 2639–2642.

Gupta, S., & Moulik, S. P. (2008). Biocompatible microemulsions and their prospective uses in drug delivery. *Journal of Pharmaceutical Sciences, 97*(1), 22–45.

Hancock, J., Callow, P., Serçe, S., Hanson, E., & Beaudry, R. (2008). Effect of cultivar, controlled atmosphere storage, and fruit ripeness on the long-term storage of highbush blueberries. *HortTechnology, 18*(2), 199–205.

Hassan, G. K., Massanet-Nicolau, J., Dinsdale, R., Jones, R. J., Abo-Aly, M. M., El-Gohary, F. A., & Guwy, A. (2019). A novel method for increasing biohydrogen production from food waste using electrodialysis. *International Journal of Hydrogen Energy, 44*(29), 14715–14720.

He, J., Zhang, L., Xu, L., Kong, F., & Xu, Z. (2020). Development of nanozyme-labeled biomimetic immunoassay for determination of sulfadiazine residue in foods. *Advances in Polymer Technology, 2020*, 1–8.

Herremans, E., Melado-Herreros, A., Defraeye, T., Verlinden, B., Hertog, M., Verboven, P., . . . & Nicolaï, B. M. (2014). Comparison of X-ray CT and MRI of watercore disorder of different apple cultivars. *Postharvest Biology and Technology, 87*, 42–50.

Houldsworth, D. W. (1980). Demineralization of whey by means of ion exchange and electrodialysis. *International Journal of Dairy Technology, 33*(2), 45–51.

Jalgaonkar, K., Mahawar, M. K., Vishwakarma, R. K., Shivhare, U. S., & Nambi, V. E. (2020). Optimization of process condition for preparation of sapota bar using refractance window drying method. *Drying Technology, 38*(3), 269–278.

Joshi, S. G., Cooper, M., Yost, A., Paff, M., Ercan, U. K., Fridman, G., . . . & Brooks, A. D. (2011). Nonthermal dielectric-barrier discharge plasma-induced inactivation involves oxidative DNA damage and membrane lipid peroxidation in Escherichia coli. *Antimicrobial Agents and Chemotherapy, 55*(3), 1053–1062.

Kim, D. I., Gwak, G., Zhan, M., & Hong, S. (2019). Sustainable dewatering of grapefruit juice through forward osmosis: Improving membrane performance, fouling control, and product quality. *Journal of Membrane Science, 578*, 53–60.

Kirtil, E., Cikrikci, S., McCarthy, M. J., & Oztop, M. H. (2017). Recent advances in time domain NMR & MRI sensors and their food applications. *Current Opinion in Food Science, 17*, 9–15.

Kumar, V., Guleria, P., & Mehta, S. K. (2017). Nanosensors for food quality and safety assessment. *Environmental Chemistry Letters, 15*(2), 165–177.

Lamagna, A., Reich, S., Rodriguez, D., & Scoccola, N. N. (2004). Performance of an e-nose in hops classification. *Sensors and Actuators B: Chemical, 102*(2), 278–283.

Laroussi, M., Lu, X., & Malott, C. M. (2003). A non-equilibrium diffuse discharge in atmospheric pressure air. *Plasma Sources Science and Technology, 12*(1), 53–56.

Longobardi, F., Casiello, G., Centonze, V., Catucci, L., & Agostiano, A. (2019). Electronic nose in combination with chemometrics for characterization of geographical origin and agronomic practices of table grape. *Food Analytical Methods, 12*(5), 1229–1237.

Mason, T. G., Wilking, J. N., Meleson, K., Chang, C. B., & Graves, S. M. (2006). Nanoemulsions: Formation, structure, and physical properties. *Journal of Physics: Condensed Matter, 18*(41), R635–R666.

Milczarek, R. R., Olsen, C. W., & Sedej, I. (2020). Quality of watermelon juice concentrated by forward osmosis and conventional processes. *Processes, 8*(12), 1568.

Milczarek, R. R., & Sedej, I. (2021). Aroma profiling of forward-osmosis watermelon juice concentrate and comparison to fresh fruit and thermal concentrate. *LWT, 151*, 112147.

Miller, G., Lowrey, N., & Senjen, R. (2008). Out of the laboratory and on to our plates: Nanotechnology in food & agriculture. *Friends of the Earth, Australia, Europe and U.S.A, 2nd edition*, 1–63.

Min, S. C., Roh, S. H., Niemira, B. A., Boyd, G., Sites, J. E., Uknalis, J., & Fan, X. (2017). In-package inhibition of E. coli O157:H7 on bulk romaine lettuce using cold plasma. *Food Microbiology, 65*, 1–6.

Monziols, M., Collewet, G., Bonneau, M., Mariette, F., Davenel, A., & Kouba, M. (2006). Quantification of muscle, subcutaneous fat and intermuscular fat in pig carcasses and cuts by magnetic resonance imaging. *Meat Science, 72*(1), 146–154.

Moreau, M., Orange, N., & Feuilloley, M. G. J. (2008). Non-thermal plasma technologies: New tools for bio-decontamination. *Biotechnology Advances, 26*(6), 610–617.

Musse, M., Quellec, S., Cambert, M., Devaux, M. F., Lahaye, M., & Mariette, F. (2009). Monitoring the postharvest ripening of tomato fruit using quantitative MRI and NMR relaxometry. *Postharvest Biology and Technology, 53*(1–2), 22–35.

Nayak, C. A., & Rastogi, N. K. (2010). Forward osmosis for the concentration of anthocyanin from Garcinia indica Choisy. *Separation and Purification Technology, 71*(2), 144–151.

Nickols-Richardson, S. M., & Piehowski, K. E. (2008). Nanotechnology in nutritional sciences. *Minerva Biotecnologica, 20*(3), 117.

Oh, Y. J., Song, A. Y., & Min, S. C. (2017). Inhibition of Salmonella typhimurium on radish sprouts using nitrogen-cold plasma. *International Journal of Food Microbiology, 249*, 66–71.

Padhi, S., & Dwivedi, M. (2022). Physico-chemical, structural, functional and powder flow properties of unripe green banana flour after the application of refractance window drying. *Future Foods, 5*, 100101.

Paixão, L. M. N., Fonteles, T. V., Oliveira, V. S., Fernandes, F. A. N., & Rodrigues, S. (2019). Cold plasma effects on functional compounds of siriguela juice. *Food and Bioprocess Technology, 12*(1), 110–121.

Pandiselvam, R., Mayookha, V. P., Kothakota, A., Ramesh, S. V., Thirumdas, R., & Juvvi, P. (2020). Biospeckle laser technique – A novel non-destructive approach for food quality and safety detection. *Trends in Food Science and Technology, 97*, 1–13.

Pelletier, J. (1992). Sterilization by the plasma procedure. *Agressologie: Revue Internationale de Physio-Biologie et de Pharmacologie Appliquées aux Effets de l'Agression, 33*(2), 105–110.

Pérez, A., Andrés, L. J., Álvarez, R., Coca, J., & Hill Jr, C. G. (1994). Electrodialysis of whey permeates and retentates obtained by ultrafiltration. *Journal of Food Process Engineering, 17*(2), 177–190.

Puligundla, P., Kim, J. W., & Mok, C. (2017). Effect of corona discharge plasma jet treatment on decontamination and sprouting of rapeseed (Brassica napus L.) seeds. *Food Control, 71*, 376–382.

Puligundla, P., Lee, T., & Mok, C. (2016). Inactivation effect of dielectric barrier discharge plasma against foodborne pathogens on the surfaces of different packaging materials. *Innovative Food Science and Emerging Technologies, 36*, 221–227.

Rahimah, S., Arbaia, R., Andoyo, R., Lembong, E., & Setiawati, T. A. (2020, February). Effect of particle size on sensory attributes of sapodilla tea (Manilkara zapota) using E-tongue. *IOP Conference Series: Earth and Environmental Science*. IOP Publishing, *443*(1), 012101. doi:10.1088/1755-1315/443/1/012101.

Rasekh, M., & Karami, H. (2021). E–nose coupled with an artificial neural network to detection of fraud in pure and industrial fruit juices. *International Journal of Food Properties, 24*(1), 592–602.

Rastogi, N. K. (2016). Opportunities and challenges in application of forward osmosis in food processing. *Critical Reviews in Food Science and Nutrition, 56*(2), 266–291.

Ravichandran, R., & Ekambaram, N. (2018). Assessment of factors influencing the concentration of betacyanin from Opuntia ficus-indica using forward osmosis: Concentration of betacyanin using forward osmosis. *Journal of Food Science and Technology, 55*(7), 2361–2369.

Sant'Anna, V., Marczak, L. D. F., & Tessaro, I. C. (2012). Membrane concentration of liquid foods by forward osmosis: Process and quality view. *Journal of Food Engineering, 111*(3), 483–489.

Santana, T. C., Silva, R. A. B. D., Pandorfi, H., Silva, M. Vd., Rodrigues, S., Guiselini, C., . . . & Gomes, N. F. (2021). Biospeckle Laser technique for mechanical damage assessment in Tommy Atkins mango fruits. *Brazilian Journal of Food Technology, 24.*

Serre, E., Rozoy, E., Pedneault, K., Lacour, S., & Bazinet, L. (2016). Deacidification of cranberry juice by electrodialysis: Impact of membrane types and configurations on acid migration and juice physicochemical characteristics. *Separation and Purification Technology, 163,* 228–237.

Silva, S. H., Lago, A. M. T., Rivera, F. P., Prado, M. E. T., Braga, R. A., & de Resende, J. V. (2018). Measurement of water activities of foods at different temperatures using biospeckle laser. *Journal of Food Measurement and Characterization, 12*(3), 2230–2239.

Song, H. P., Litchfield, J. B., & Morris, H. D. (1992). Three-dimensional microscopic MRI of maize kernels during drying. *Journal of Agricultural Engineering Research, 53,* 51–69.

Song, Y., Zang, X., Kamal, T., Bi, J., Cong, S., Zhu, B., & Tan, M. (2017). Real-time detection of water dynamics in abalone (Haliotis discus hannai Ino) during drying and rehydration processes assessed by LF-NMR and MRI. *Drying Technology, 36*(1), 72–83.

Surowsky, B., Fischer, A., Schlueter, O., & Knorr, D. (2013). Cold plasma effects on enzyme activity in a model food system. *Innovative Food Science and Emerging Technologies, 19,* 146–152.

Tazi, I., Choiriyah, A., Siswanta, D., & Triyana, K. (2017). Detection of taste change of bovine and goat milk in room ambient using electronic tongue. *Indonesian Journal of Chemistry, 17*(3), 422–430.

Tendero, C., Tixier, C., Tristant, P., Desmaison, J., & Leprince, P. (2006). Atmospheric pressure plasmas: A review. *Spectrochimica Acta Part B: Atomic Spectroscopy, 61*(1), 2–30.

Thazin, Y., Pobkrut, T., & Kerdcharoen, T. (2018, January). Prediction of acidity levels of fresh roasted coffees using e-nose and artificial neural network. In *2018 10th International Conference on Knowledge and Smart Technology (KST)* (pp. 210–215). IEEE.

Thybo, A. K., Szczypiński, P. M., Karlsson, A. H., Dønstrup, S., Stødkilde-Jørgensen, H. S., & Andersen, H. J. (2004). Prediction of sensory texture quality attributes of cooked potatoes by NMR-imaging (MRI) of raw potatoes in combination with different image analysis methods. *Journal of Food Engineering, 61*(1), 91–100.

Toepfl, S., Mathys, A., Heinz, V., & Knorr, D. (2006). Review: Potential of high hydrostatic pressure and pulsed electric fields for energy efficient and environmentally friendly food processing. *Food Reviews International, 22*(4), 405–423.

Topuz, A., Dincer, C., Özdemir, K. S., Feng, H., & Kushad, M. (2011). Influence of different drying methods on carotenoids and capsaicinoids of paprika (Cv., Jalapeno). *Food Chemistry, 129*(3), 860–865.

Torlak, E., & Sert, D. (2013). Inactivation of Cronobacter by gaseous ozone in milk powders with different fat contents. *International Dairy Journal, 32*(2), 121–125.

Trombete, F. M., Porto, Y. D., Freitas-Silva, O., Pereira, R. V., Direito, G. M., Saldanha, T., & Fraga, M. E. (2017). Efficacy of ozone treatment on mycotoxins and fungal reduction in artificially contaminated soft wheat grains. *Journal of Food Processing and Preservation, 41*(3), e12927.

Vera, E., Sandeaux, J., Persin, F., Pourcelly, G., Dornier, M., & Ruales, J. (2009). Deacidification of passion fruit juice by electrodialysis with bipolar membrane after different pretreatments. *Journal of Food Engineering, 90*(1), 67–73.

Wang, T. M., & Rogers, M. A. (2015). Biomimicry – An approach to engineering oils into solid fats. *Lipid Technology, 27*(8), 175–178.

Weber, C., Dai Pra, A. L., Passoni, L. I., Rabal, H. J., Trivi, M., & Poggio Aguerre, G. J. (2014). Determination of maize hardness by biospeckle and fuzzy granularity. *Food Science and Nutrition, 2*(5), 557–564.

Wu, J., Doan, H., & Cuenca, M. A. (2006). Investigation of gaseous ozone as an anti-fungal fumigant for stored wheat. *Journal of Chemical Technology & Biotechnology: International Research in Process, Environmental & Clean Technology, 81*(7), 1288–1293.

Xu, Y., Peng, X., Tang, C. Y., Fu, Q. S., & Nie, S. (2010). Effect of draw solution concentration and operating conditions on forward osmosis and pressure retarded osmosis performance in a spiral wound module. *Journal of Membrane Science, 348*(1–2), 298–309.

Zheng, L., Cai, G., Qi, W., Wang, S., Wang, M., & Lin, J. (2019). Optical biosensor for rapid detection of Salmonella typhimurium based on porous gold@ platinum nanocatalysts and a 3D fluidic chip. *ACS sensors, 5*(1), 65–72.

Chapter 2

Electrodialysis

A Novel Technology in the Food Industry

Isha Dudeja, Priyanka, Pooja Nikhanj, and Arashdeep Singh

2.1 Introduction

Electrodialysis (ED) is an electro-membrane method which modifies the ionic composition of liquids by using ion permeable membranes and direct electrical potential (Mondor *et al.* 2012). According to International Union of Pure and Applied Chemistry (IUPAC), ED is a "membrane-based separation technique in which ions are made to pass through an ion-selective membrane under the influence of an electric field" (Koros *et al.* 1996; Ronald 1987; Grebenyuk and Grebenyuk 2002). Over the past century, ED has transformed from a lab curiosity to a potent instrument with several industrial applications. The application of different ED techniques to challenging separation operations in several food and beverage industries is particularly intriguing. When it comes to separation, the dairy, wine, juice, and sugar sectors frequently use ED. In addition to being a technique for desalination, ED in all of its forms has emerged as a practical and affordable solution to challenges in the dairy, wine, and juice sectors (Hestekin *et al.* 2010). The agrifood sectors are using ED and a number of other membrane-based separation methods more frequently. They are employed to enhance, purify, or alter food. Some of the factors favouring this

DOI: 10.1201/9781003217138-2

technology are low energy consumption, efficiency, modular design, and convenience of operation, as well as the heat sensitivity of many food products. Despite the fact that ED is a membrane separation process, it has some variations from other methods like nanofiltration, reverse osmosis, and ultrafiltration because it separates particles based on their electrical charges rather than their size (Bazinet 1998). The theory and applications of ED as it relates to the food industry will be covered in this chapter.

2.2 History of Electrodialysis

Ostwald conducted the first ion-exchange membrane experiments in the early 1890s, which greatly expanded the possibilities for membrane separation technology (1890). The information of membrane potential and the phenomenon of Donnan exclusion were established few years later (Donnan 1995). Manegold and Kalauch (1939) were the first to propose the idea of ED. For separation of ions from water, they set up an arrangement of ion exchange membranes with unlike charge affinity, that is, cationic as well as anionic ion exchange membranes. The following year, the setup Meyer and Strauss (1940) was updated by and expanded to combine numerous similar pairs of membranes into a multi-cell plan between a single pair of electrodes. Newly created polymer membranes and this configuration quickly made ED the preferred technology for industrial desalination plants. ED reversal (EDR), a version of this method, was introduced in the 1970s to solve some inherent issues in conventional ED (Grebenyuk and Grebenyuk 2002; Hughes 1992). EDR replaced conventional ED in desalination applications because it had lower operating expenses, especially when it came to maintaining membrane systems. The application of ED to other commercial separation issues was made easier by the lower operating costs associated with EDR. The invention of bipolar membranes in 1977 and perfluoro based membranes in 1979, increased the range of industrial separation for which ED technology could be used (Nagasubramanian et al. 1977).

Some clarifications in the early history of ED are necessary for factual accuracy (Shaposhnik and Kesore 1997). Early studies (Maigrot and Sabates 1980; Cassel and Kempe 1894) with non-selective membranes are not recognized as ED related, despite the fact that some of them (Schollmeyer 1902) used the name "ED". The first ion-selective membranes were produced in 1911 and 1939. However, ED theory existed much earlier (Donnan 1995). Then, in 1939 (Manegold and Kalauch), the concept for the first multicell ED was created. The first artificial ion-exchange membranes were first created in 1950 by Juda and McRae. The same year, Kressman created his membranes (1950). His membranes were modified natural products—impregnated parchment membranes—rather than manufactured ones. Thus, the development of ED technology as we know it today began in 1950 (Grebenyuk and Grebenyuk 2002).

2.3 Principles of Electrodialysis

In general ED system flow, compartments or cells are composed of ion exchange membranes arranged systematically, which separates the ions under the influence of electric fields. Sets of anion (A) and cation (C) exchange membranes organized in an alternating way vertically between an anode and a cathode to form individual cells are shown in Figure 2.1, demonstrating the principle of ED. Application of direct current (DC) voltage aids in the movement of ions through the ion exchange membrane during the electrochemical separation process or ED (Hestekin 2010). An ionic solution causes the migration of cations (positively charged) and anions (negatively charged) towards the cathode and anode, respectively. For instance, the separation of ions in an aqueous saline solution during transport via cells in an ED system. Based on the affinity of membranes, anion exchange membranes attract cations but not anions, while cation exchange membranes trap anions and facilitate the movement of cations. As already stated, the electrical potential generated between the electrodes drives ion movement during ED (Strathmann 1995). Ion concentration gradients work together with electric potential gradients to drive the movement of ions in ED devices. In ED systems, the electrical potential gradients produce forces that are substantially greater than the forces produced by ion concentration gradients. Even though the transportation of ions take place in opposite directions while the ED system is in operation, the amount of electrical current transported may or may not be equal. The transport number for certain types of ion is generally defined as the percentage of the electric current on anions or cations; in ED systems, the total transport number is 1. The cation transport number is determined by the speed of cations in the outwardly applied electric field, and the anion transport number is determined by the speed of the anions in the same direction. Because ions vary in size and charge, different ionic species have different ion transport numbers. In ED and other membrane parting technologies, the transport number of ions with the same charge as the ion exchange membranes is zero. The transport number is 1 for ions with opposite charges to the charged group of the membrane. When the transport number approaches 1, ions can pass across the membrane; when it is close to 0, ions cannot. With ion exchange membranes, separation can be accomplished via a difference in transport number (Hestekin 2010). Overall, this causes alternating compartments to have higher ion concentrations while concurrently depleting the other compartments. The intense solution is sometimes known as concentrate or brine, and depleted or diluted solution is typically called the diluent. Separating the relevant components can be accomplished by adjusting the salt and ion concentrations in ED systems. One of the most popular types of ED used for desalination and deionization

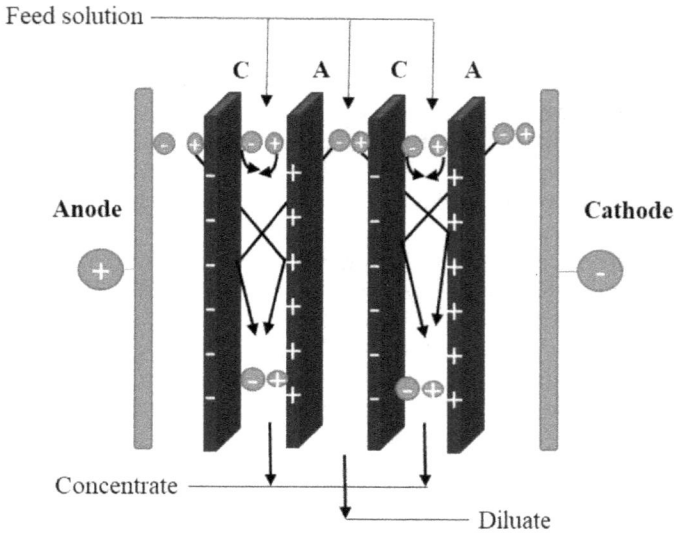

Figure 2.1 Electrodialysis unit with cell pair.

is seen in Figure 2.1. However, there are a variety of additional techniques that are quite similar to standard ED, and they use diverse topologies of neutral or ion-exchange membranes with or exclusive of an electrical potential driving force (Strathmann 1991). Donnan dialysis, diffusion dialysis, isoelectro-dialytic focusing, and electro-dialytic water dissociation are some of the procedures that are used. The ability of ED to separate ions from a particular mixture of other molecules—a process known as mass separation—depends heavily on the quality features of the membranes used in the device. The running cost of the process, which is primarily driven by energy utilization and capital expenditures for a plant with the specified capability—determines its profitability. Membrane characteristics influence energy use and investment expenses to a great extent, but other process design factors like flow rates, current densities, and cell sizes also have an impact (Strathmann 1995)

2.4 Parts of ED systems

The design of ED systems is generally application oriented. The design of the membrane stack determines how many ions need to be eliminated. A membrane stack may be arranged vertically or horizontally.

2.4.1 Cell Pairs

The fundamental components of an ED membrane stack are cell pairs (Figure 2.1). Each built stack contains two electrodes as well as groupings of cell pair groups. A cell pair has the combined region occupied by a dilute solution between two adjacent membranes, a concentrated solution between two adjacent membranes, and two adjacent cation and anion-exchange membranes. In an ED stack, a cell pair is a repeating unit. There could be hundreds of these membranes in a practical ED stack (Schaffer and Mintz 1966). The combined volume of the concentrated solution between the two adjacent membranes, the dilute solution between the two adjacent membranes, and the two contiguous anion and cation exchange membranes make up a cell pair. Stacks of over 600 cell pairs can be constructed for industrial purposes. The quality of the source water mostly determines the number of cell pairs required to attain the particular product's water quality (Strathmann 1995, 2004, 2005).

Water flow in sections:

1. Unlike the concentrate stream, which only flows analogous during concentrating portions, the source water (feed) only flows through de-mineralizing regions.
2. As feed water passes through the membranes, ions are electrically transferred from the de-mineralized stream to the concentrate stream.
3. Other streams are not affected by the two electrode compartmental flows. A degasifier is used to expel reaction gases from the electrode waste stream.
4. The membranes and spacers are compressed by steel block top and bottom plates to prevent leakage inside the stack.

In a batch ED system, source water is repeatedly cycled through de-mineralizing spacers in a single membrane stack and then returned to the holding tank until the necessary purity is reached. The manufacture rate is affected by the concentration of dissolved minerals in the source water as well as the level of demineralization required. Re-circulating the concentrate stream reduces the volume of wastewater even further, and acid must be provided regularly to minimize membrane stack scaling. The unidirectional continuous-type ED system was the second kind of commercially available system. Each level in this system's membrane stack, which has two stages in total, aids in de-mineralizing the water. The de-mineralized stream enters the stack only once before exiting as product water. Acid is introduced into the concentrate stream to avoid scaling and is partially recycled to reduce effluent volume (Valero *et al.* 2011).

2.4.2 Membranes

A membrane is a permselective barrier that partitions two phases (Mulder 1996). A driving force can cause some substances to permeate the membrane while keeping others in place. As a result, a membrane has the ability to move specific parts from one phase to another. There are numerous distinct types of membranes that can be divided into groups based on characteristics such as their driving force, morphology (porous, non-porous, asymmetric, symmetric, neutral, or charged), and others (concentration difference, pressure difference, electrical potential difference) (Krol 1999). Membranes are created as thin foils with ion exchange groups attached and are made of tiny polymer particles that are anchored by a polymer matrix. Membranes are strengthened with synthetic fibre so that they are impermeable to water under pressure and to increase their mechanical strength (AWWA 1995). Ion-exchange membranes applied in ED are films made from ion-exchange resins, so they are made of extremely swollen gels with permanent negative or positive charges. Because they may adsorb varying amounts of water depending on pH, electrolyte concentration, and temperature, ion exchange membranes are often weak and dimensionally unstable, whereas ion exchange membranes have greater mechanical force and flexibility thanks to the addition of the right reinforcing elements. Two main groups of commercially accessible ion exchange membranes are homogeneous and heterogeneous (Strathmann 1995). Either the functionalization of a polymer film through polystyrene film sulphonation or the polymerization of functional monomers (such as the polycondensation of phenol or phenol-sulphonic acid with formaldehyde) is used to create homogeneous ion exchange membranes. Heterogeneous ion exchange membranes are made by dry ion exchange membrane resin dispersion in a solution or in a melted polymer matrix or by melting and pressing dry ion exchange resins with a granulated polymer. There are two general types of membranes employed for ED. Thin anion or cation exchange resin sheets or films laminated with synthetic textiles are utilized as ED membranes because they provide the requisite mechanical strength. Most commercially available ion-exchange membranes are made of hydrophobic polymers like polyethylene, polystyrene, or polysulfone that have been crosslinked with divinylbenzene. The level of crosslinking affects how much a polymer swells and increases the mechanical strength of a membrane. The mechanical, chemical, and thermal stability of the membranes is largely a function of the material from which they are produced. The crosslinking level causes an increase in the membrane's selectivity, stability and electrical resistance. Electrical resistance and membrane perm-selectivity are both influenced by the type, concentration, and density of permanent charges on the polymer (Mondor *et al.* 2012).

2.4.2.1 Monopolar Membranes

Monopolar membranes have an affinity for one kind of charged ion (Fidaleo and Moresi 2006). Basically, two different types of ion exchange membranes are present in ED systems:

Cation exchange membranes: Cation exchange membranes are electrically conductive in nature and have high affinity for positively charged ions, thus allowing only these ions to pass through them. The fixed charge sulfonic acid group (SO_3^-) and the carboxylic acid group (COO^-) are typically applied to cationic membranes where SO_3^- nearly dissociates over the whole pH range and COO^- essentially does not dissociate in the pH range of 3 (Strathmann 2004; Mondor *et al.* 2012). Most commercially available cation membranes are constructed by crosslinking sulfonated polystyrene, $-SO_3H$ groups in sulfonated polystyrene splits in water to give H^+ and SO_3^- a mobile and constant charge, respectively (Valero *et al.* 2011).

Anion exchange membranes: Anion change membranes are also electrically conductive in nature and enable the passage of negatively charged ions exclusively. There are two types of fixed charges on these membranes: NR_3^+ and NH_2R^+, that is, a quaternary ammonium group and secondary ammonium group, respectively, where former is entirely dissociated across the entire pH range, and the latter is only marginally dissociated. The mechanical stability and perm-selectivity properties of these membranes are exaggerated by the concentration or strength of permanent charges due to the effect on their swelling power or degree of swelling (Strathmann 2004; Mondor *et al.* 2012). Since quaternary ammonium groups ($-NR_3^+$ OH^-) have fixed positive charges in the membrane matrix, they typically resist positive ions (Valero *et al.* 2011).

Each membrane varies depending on the manufacturer, but they typically range from 0.1–0.6 mm thickness and can be homogenous or heterogeneous depending on how charge groups are connected to the matrix or their chemical makeup (Xu 2005). Both types of membranes share characteristics such as insolubility in water-based solutions, lower electrical resistance, appreciable rigidity for management during stack assembly, resistance to pH changes between 1 and 10, operation temperatures above

46°C, resistance to fouling, osmotic swelling, great shell life, and hand washability. The ability of the membranes to differentiate between various ions in order to permit passage or permeation through the membrane is known as perm-selectivity (or ion selection). In this respect, membranes may be designed to prevent the passage of divalent cations or anions like calcium, magnesium, or sulphates. Some membranes have strong permeability and high transport numbers for monovalent anions like Cl^- or NO_3^- but poor transport numbers and relatively low penetration rates for divalent (SO_4^{2-}) or trivalent ions (PO_4^{3-}) or analogous anions. The anion membrane is given a particular treatment to do this, and the effect can be used to segregate different ions. The monovalent anion membrane exhibits the highest specificity (Xu 2005). Charged groups are attached physically to the matrix membrane in heterogeneous membranes as opposed to homogeneous membranes, where they are chemically bound. Using a few drops of 0.05% dye solutions such as methyl orange/methylene blue, cationic and anionic exchange membranes can easily be distinguished as they will be stained a deep blue or yellow colour, respectively. The characteristics of these membranes heavily influence both the functional and economic viability of the industrial framework. Important electrodialytic features that must be taken into account include electrical resistance, permanent selectivity, ion-exchange capacity, superior mechanical and form stability, solvent transfer, and chemical tolerance. Ion-exchange membranes have a variety of features, many of which are determined by variables that frequently contradict one another. The electrical resistance of these membranes is an essential ability of ED that is required for it to be used economically. The capacity of ion mobility and ion exchange across the membrane decides the electrical resistance of the membrane that ultimately determines the energy requirements of the ED process. The membrane matrix's fixed ionic charges dictate the electrical resistance; for example, low electrical resistance results from a higher concentration of fixed charges, ultimately causing significant swelling and poor mechanical stability. The extent of cross-linking can be increased to increase the mechanical strength and electrical resistance of membrane (Strathmann 2004). The perm-selectivity of a membrane is defined as its capacity to permit the transport of counter-ions while inhibiting the crossing of co-ions. It is an essential attribute compared to other properties of membrane such as their ion exchange capacity that is defined as the number of fixed charges per dry mass of the membrane (Young 1974). The concentration of the co-ions in the membrane is rather low, and counter-ions are thought to carry the majority of the electric current in a membrane. High membrane perm-selectivity is vital since it directly affects the efficiency of the process. The highest concentration of the concentrate that may be achieved will be impacted by solvent transfer by osmosis and electroosmosis. Electroosmosis is a phenomenon where the solvent is moved across the membrane under the influence of an electrical current applied in the hydration ion shelf. Osmosis, which involves the transfer of water from a diluted to a concentrated solution, can also take place in the absence of an electrical field (Mondor *et al.* 2012).

2.4.2.2 Bipolar Membrane

The bipolar membrane, or bpm, was one modern form of electro-membrane that began to be commercialized towards the end of the 1980s (Bazinet *et al* 1998). It consists of a physically or chemically bound anion- and cation-exchange membrane and a very lean hydrophilic layer (less than 5 nm), where water molecules diffuse from the surrounding aqueous salt solutions. These molecules split up into hydrogen (H^+) and hydroxyl (OH^-) ions in the existence of an electrical field (Mani 1991). If the bipolar membrane is correctly oriented (no current reversal is permitted in water splitting), these ions can migrate out of that layer. OH^- and H^+ ion flows across the anion exchange face the anode, and the cation-exchange membrane faces the cathode. Due to the conditions, bipolar membranes effectively generate and concentrate hydrogen and hydroxyl ions (up to 10 kmol/m³) at their surfaces, producing basic and acidic solutions, respectively. The primary benefits of bipolar membranes include the absence of gas formation at their surfaces or within the bipolar membranes themselves, a power requirement for dissociating water into O_2 and H_2 that is about half that of electrolytic cells, the minimal production of waste streams when using dilute acids or bases (1 kmol/m³), and reduced downstream purification systems. A bipolar membrane must have some essential qualities, including low resistance at high current densities, great ion specificity, considerable water dissociation rates, low co-ion transport rates, excellent chemical resistance, and thermal durability in strong acids and bases. Higher concentrations of fixed charges in the polymer such as $-SO_3H$ and NR_3^+/NH_2R^+ groups in the cation and anion exchange membrane, respectively, certainly decrease the electrical resistance of the cationic and anionic layers of bipolar membranes (Fidaleo and Moresi 2006). Bipolar membrane-enabled ED stacks can be set up in two or three compartments and run in the same single- or multi-passage continuous or batch mode as any other traditional ED stack. A salt solution is divided into an acidic and basic stream using the most popular three-compartment configuration, which is created by sandwiching a bpm between an anionic (a) and cationic (c) membrane. Each stream exits the compartments that are restricted by a bipolar and anionic membrane and a bipolar and cationic membrane, respectively (Figure 2.2). In this structure, up to 200 cells can be arranged in a single stack that is equipped with a system of manifolds that can simultaneously feed the three related compartments, the acid, base, and dilute, in addition to the two electrode compartments. It is feasible to separate a salt of an organic acid into the free acid and the equivalent hydroxide by adding a salt solution to the diluting

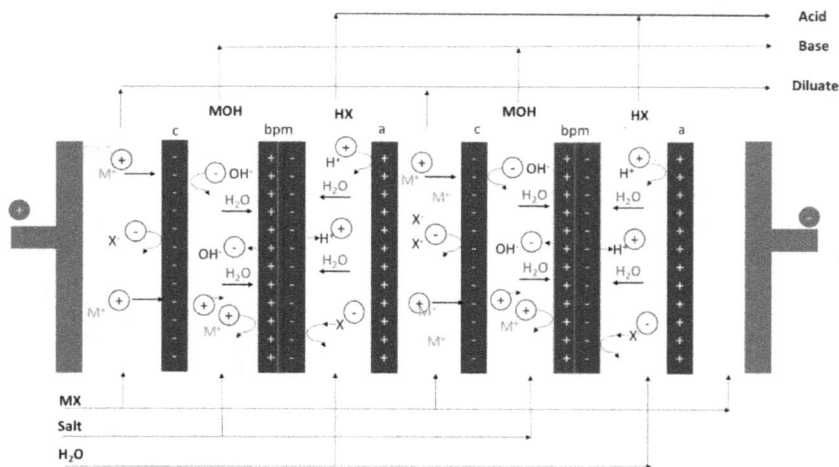

Figure 2.2 Bipolar membrane.

compartments, water to both the acid and base compartments, and a DC across the electrodes. Only 1–2% of the total power used in this process is used at the electrodes, where small amounts of hydrogen and oxygen are produced. The overall investment and operating costs can be reduced by coupling a bpm with only a cation- or anion-exchange membrane in a two-compartment configuration. This is because the ED unit requires fewer membranes to be installed and replaced, fewer process loops, and less power to operate under constant-current density. The sodium salts of acetic, lactic, and formic acids as well as sodium glycinate can be converted into a concentrated stream that contains the corresponding strong base or acid and a diluting stream that contains the residual fraction of the salt (e.g., 1–2% w/w) along with the corresponding strong base or acid using two-compartment cells made of bipolar and cation- or anion-exchange membranes. Due to the weak electric conductivity of the corresponding weakly dissociated acid or base, it is impracticable to use a three-compartment cell in these circumstances. Bipolar membrane-enabled ED is considered the state of the art of ED because it combines the functions of ED and water splitting of bipolar membranes (Huang and Klein Xu 2006). As seen in Figure 2.2, bipolar membranes (bpm), cation (c), and anion (a) exchange membranes are alternately positioned between the anode and cathode in a typical stack bipolar membrane-enabled ED (c-bpm-a-c-bpm-a). The four cell compartments that are produced after this arrangement of membranes are acid, base, salt, and electrode compartments. As electrical potential difference is generated, anions move toward the anode, cross the "a" membrane, and then mix with

H[+] produced by the "bpm". This results in production of acid by acid compartments. Cations move in the opposite direction and cross the "c" membrane, then associate with the OH[-] produced by "bpm". Salt compartment ionic component depletion takes place similarly to how it does throughout the ED process. As a result, salts can be converted into their corresponding compounds. Bipolar membranes have several advantages over solvent electrolysis for producing alkali, including no gas or by-product formation, a smaller voltage drop, maximum energy efficiency, space savings, and ease of installation and operation (Xu 2005). As a result, this method finds extensive use in food processing, pollution management, and chemical and biochemical synthesis. Bipolar membrane-enabled ED is emerging as a very favourable alternative to conventional acidifying or basification technologies used in the food processing industry. Lower heat damage to the product, economical energy usage, and inexpensive equipment are the benefits of BMED over conventional techniques. The use of IE membranes improves the process's energy efficiency and cleanliness while also raising the added value of the finished goods (Wang *et al.* 2019).

Bipolar membranes were used for the first time in the industry to recover bases and acids from salt streams (Mani 1991; Nagasubarmanian *et al.* 1977). The water-splitting process results from the integration of the advantages of combining conventional ion exchange membranes with bipolar membranes in a cell stack. An aqueous NaCl solution is transformed into an acid and (HCl) base (NaOH) when it is supplied to the stack along with a direct current to the electrodes. As well as organic salts, other salts such as $KF, KNO_3, NH_4Cl, Na_2SO_4$, and KCl can also be separated. The first industrial bipolar membrane facility at Washington Steel in Pennsylvania was established in 1988 (Bazinet 1998); complete acid recovery is made possible by the renewal of HF and HNO_3 acids from stainless-steel pickling liquid. This approach was employed by Boyaval *et al.* (1993), and Alvarez *et al.* (1997) to produce a few concentrated acids such as lactic, propionic, and salicylic acids at laboratory level, respectively. With the application of ED in a cell with membrane pair, that is, a bipolar membrane and an ion-exchange membrane, the acidity of an edible liquid (sugar syrups, sauces, wines, juices, fruit and vegetable and tomato paste) can be altered. This latter use paves the way for novel uses of bipolar membranes that go ahead of the creation of acids. Bipolar membrane ED appears to be an intriguing ED application for the food business. Other food products whose chemical, enzymatic, and microbiological stability is significantly pH dependent can be treated using the change in pH obtained in an ED cell with a bipolar membrane. On the other hand, high-purity protein fractions can also be separated using bipolar membrane electrodialysis (BMED). Since the rates of electro-alkalinization and electro-acidification are controlled by the efficient current density in the cell, the process may be precisely controlled. However, to reduce the overall resistance of the ED cell, the products to be treated by bipolar membrane-enabled ED must have a low viscosity to circulate readily in an ED cell and a relatively high mineral content. The high equipment cost and bipolar membrane-enabled ED's propensity for blockage are some of its main drawbacks (VanNispen 1991). These drawbacks, meanwhile, are a result of this technology's uniqueness. The advancement of this technology and its uses in the food sector will contribute to lowering the price of bipolar membranes and ED cells. Additionally, the blocking of soybean protein in BMED is primarily caused by spacer fouling, which results from an ED cell with a non-optimized hydrodynamic design rather than membrane fouling, as in ultrafiltration and nanofiltration. At higher concentrations, the particles in suspension cause the fouling to clump together in the spacers; however, a centrifuge may enable the recovery of the particles and lessen fouling (Bazinet *et al.* 1998).

2.4.3 Spacers

In order to form independent flow channels and manifold all the demineralized streams and concentrate streams together, spacers, made up of polypropylene/low-density polyethylene, are positioned between membranes in alternating fashion in the stack. By turning a similar spacer 180 degrees, demineralizing and concentrating spacers can be made. Concentrating spacers work by stopping the concentrate stream from entering the demineralized stream, whereas demineralizing spacers permit water to pass through membrane surfaces for ion removal. The liquid flow from a spacer with a "tortuous path", where the spacer is twisted back on itself, is substantially longer than the unit's linear dimensions. A "sheet flow", which consists of a plastic screen with an open frame that divides the membranes, is another type of spacer. Spacers are operated at lower flow velocity in order to obtain a level of desalting in each transport cycle through the stack comparable to tortuous path or sheet flow spacers. In general, an increase in turbulence encourages water mixing, usage of the membrane surface, and ion transport. In addition to dispersing particles or slime from the membrane surface, turbulence caused by spacers also draws ions to the membrane surface. Spacers come in a variety of styles and sizes to meet various design requirements. Different flow pathways that affect water velocity through the membrane stack and the interaction time of the source water with the membrane are the primary differences between spacer models. Water velocity is a crucial design factor when selecting a spacer since it determines how much desalination and mixing take place across membranes. The rates of flow for both streams must be balanced in EDR systems, as the same spacers are utilized for demineralized and concentrated water. This prevents excessive differential pressures across the membranes (Valero *et al.* 2011).

At cathode		At anode	
Cations (Na+) attraction		Anions (Cl⁻) attraction	

$$2H_2O \longrightarrow H_2 + 2\,OH^-$$

$$2H_2O \longrightarrow 4H^+ + O_2 + 4e^-$$

Cl_2 gas formed

pH raised↑, precipitate of $CaCO_3$↓

pH lowered↓, precipitate of $CaCO_3$ dissociates

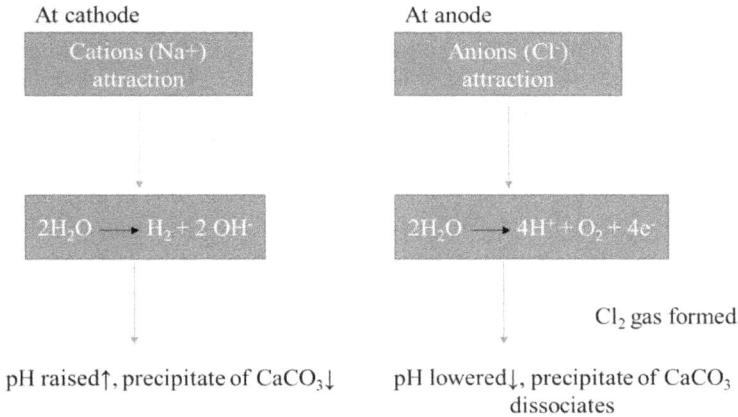

Figure 2.3 Electrode system of ED.

2.4.4 Electrodes

At either end of the stack, metal electrodes run DC current into the stack. The electrode compartment consists of an electrode, a heavy cation membrane, and an electrode water-flow spacer. Because of high thickness of electrode spacer compared to a typical spacer, water moves more quickly, preventing scaling. Additionally, these spacers keep electrode waste from getting into the stack's main flow channels (Valero *et al.* 2011). Electrodes are often constructed of titanium and coated with platinum due to the anode compartments' corrosive nature. The ionic makeup of the source water and the current applied affect how long it will last. High amperages and chlorine concentrations in the source water shorten the life of electrodes. Additionally, polarity reversal (as in EDR) drastically reduces the electrode lifetime compared to non-reversing devices (AWWA 1995). Figure 2.3 shows the reactions occurring at electrode when a DC voltage is placed across the electrodes (AWWA 1995).

2.5 Applications of Electrodialysis in Food Processing

ED is a unified effort for the separation of ions present in a solution based on their specific electromigration through semi-penetrable membranes affected by a likely slope (Lacey and Loeb 1972; Strathmann 1991). Attributable to their selectivity, ion membranes (IEMs) permit the transport of just cations (cation-exchange membranes) or anions (anion-exchange films) and in this way can be utilized to concentrate, eliminate, or separate electrolytes. Despite the fact that the first modern ED application in the food industry began in 1960 and involved demineralizing cheddar whey for use in child food sources, ED has a deeper historical background than is typically acknowledged (Shaposhnik and Kesore 1997). Maigrot and Sabates (1890) were the first to suggest a method for purifying up sugar syrup that combined electrolysis and dialysis. The primary example consisted of two carbon cathodes that were isolated by permanganate paper-based films and were driven by dynamos. Potassium, sodium, magnesium, and calcium cations would typically flow into the cathodic compartment once the sugar syrup had been poured into the focal anodic compartment and the dynamo had been turned on. It was possible to prevent sparingly soluble hydroxides from escalating by regulating the pH in this compartment using the litmus paper. The ED process was then terminated as soon as the pointer became blue.

In 1939, Manegold and Kalauch gathered a three-compartment ED device comprising a permselective anion-exchange membrane and a cation-exchange one. It was, nonetheless, not until the mid-1950s that the production of specific ion exchange membrane permitted multi-compartment electro-dialyzers to be constructed (Shaposhnik and Kesore 1997).

Today, ED is a membrane-based technique that has helped in advancements of agri-food businesses to sanitize or alter food sources. The low energy utilization, proficiency, and usability, as well as the intensity responsiveness of numerous food items, are among the explanations behind their fast development. In spite of the fact that ED is a membrane-based technique, it isolates particles according to their electrical charges, whereas other techniques like ultrafiltration, nanofiltration, and turn-around assimilation separate particles according to their sizes (Bazinet et al. 1998). ED involves an electric field—the main thrust and ion-exchange membrane to perform the separation. ED is referred to as a consolidated technique of electrolysis and dialysis (Shaposhnik and Kesore 1997).

ED can be executed with two types of primary cells: multi-layer cells for weakening fixation and water separation applications and electrolysis cells intended for oxide-decreasing reactions. In multi-layer cells, the film transport peculiarity is immediate, while electrochemical reactions occurring at the terminals do not interface with the partition interaction. The electrodes are straightforward electrical terminals drenched in electrolytes for ongoing exchange. Electrolysis cells work with one layer that isolates the two arrangements circling within each terminal section. This application depends upon the terminal redox reactions that are electrolysis-explicit properties. Anode initiates oxidation, and decreases happen over the cathode end (Klein et al. 1987).

ED is a widely demonstrated innovation, with a large number of frameworks working around the world. Electro-dialysis outperforms as a desalting interaction with entire plant limit over in Japan and Europe in order to counter assimilation and refining (Shaposhnik and Kesore 1997). The primary logical paper on ED was distributed by Morse and Penetrate in 1903 (Shaposhnik and Kesore 1997), yet the quantitative hypothesis of charged films was only accessible 50 years after the fact (Teorell 1953). It integrated the electrostatic repulsion of co-particles from the proper charges of the membrane, as anticipated by the thermodynamical hypothesis of film balance created by Donnan (1995). The current biggest area of use for ED is in the desalination process of bitter water for its conversion into consumable water (Audinos 1992; Strathmann 1991) and deashing of milk whey to get important crude materials for child food sources (Batchelder 1987). In any case, ED applications are still in their beginnings, and it is likely that this is due to the high explicit electro-membrane costs or their short lifetime (this being no longer than 1 year, particularly if the feed arrangement is fouling or ED, or alternatively ED detachment plant has not been very well planned).

The main goal of this chapter is to evaluate both existing and possible ED applications that, in the short term, have the potential of being particularly significant for the food sector area as well as the important mass vehicle circumstances that might be helpful to the ED unit plan or enhancement. Diffusion dialysis, Donnan dialysis, and bipolar membrane dialysis are a few examples of techniques that are used on a large scale in the present day and can be regarded as cutting-edge inventions. Numerous cycles, including continual electrodeionization, electrodialytic water separations using bipolar filaments, and power ageing with turned-around ED, are still being improved. In any case, interest in these processes is quickly developing, and various new applications have been distinguished. In any case, all of the presently accessible electro-membrane cycles and parts utilized in these cycles still have specialized and business impediments and regardless of significant continuous improvement, there is a need for further examination to further develop items and processes.

The current ED industry has maintained a consistent development pace of around 15% for the last 15 years (Srikanth 2004). The main modern ED application is still the production of consumable water from salt water. In any case, different applications either in the semiconductor business for the development of ultrapure or at least totally deionized water without the substance regeneration of ion exchange resins (IERs) or in the food business (i.e., whey demineralization, tartaric adjustment of wine, organic product juice deacidification, and molasses desalting) are acquiring expanding significance with enormous scope for modern establishment. Over the last 10–15 years, various advancements in membrane and framework innovation have made EDR a particularly appealing innovation both with regard to execution and economics. Numerous uses of EDR innovations could be established worldwide. Different providers and applications are involved for scaling up the technology. The desalting method is applied primarily to the salt water process, specific modern applications, and tertiary wastewater production, ranging from mining to drug and food and refreshments industries.

2.5.1 Electrodialysis in the Food and Dairy Industry

The use of ED for demineralization depends on the item to be dealt with. For example, when ED is applied to acidic whey as a pre-treatment for lactose production, the interaction is only financially suitable for demineralization up to around 50%; more significant levels require extreme utilization of electrical power and distribution time (Durham 2009). The use of ED in film filtration processes has not yet completely evolved, and improved understanding is required to advance this innovation mechanically. For instance, nanofiltration (NF) re-tentate (NFR), a side stream structure without milk lactose production, contains elevated centralization of lactose and certain minerals, resulting in a pungent aroma and severe taste when used as a supplement to different items. Subsequently, effective NFR demineralization by ED can permit working on the equilibrium of minerals and hence result in the improvement of new uses or recycling of supplements in dairy handling. No previous review has introduced the use of ED on layer filtration side streams like NFR. The goal is to determine the effectiveness of demineralization at various NFR focuses, assess the item synthesis when handling, and develop a forward-looking model for the targeted expulsion of cationic minerals in order to examine the mechanical feasibility of the ED cycle for demineralization without the production of lactose milk.

In the dairy industry, the handling of acidic whey (a by-product of cream and cheddar making) casein production, and strained yogurt result in high lactic acid fixation, causing functional issues in downstream shower drying tasks because of expanded powder tenacity. So, to overcome this, ED, a widely demonstrated demineralization innovation, was utilized to eliminate lactate particles from acidic whey. The proportion of lactic acid to lactose is lowered to a similar level as that found in sweet whey; 80% of the lactate particles need to be separate. Energy utilization (0.014 kW h/kg for whey handled) to accomplish 90% demineralization of acidic whey was equivalent to the energy necessity found for sweet whey demineralization in commonplace business ED units, showing the achievability of this method (Chen et al. 2016).

ED has a positive impact on lactic acid removal and demineralization, as the drying process of acidic whey is obstructed by its elevated mineral content and natural corrosive items, and their evacuation is performed economically through expensive and environmentally impactful sequential processes. Previous studies exhibited the capacity for eliminating these components by ED alone, but with a central issue: membrane scaling. In this method, two different types of pulsed electric field (PEF) were tried and contrasted with ordinary direct current to assess the capability of PEF to relieve film scaling and manipulate lactic acid and salt expulsions.

In the fruit juice industry, ED handling may be applied in three primary regions: de-fermentation, desalting, and enzymatic-sautéing restraint. Some acidic natural product juices must be de-fermented by expanding sugar or salt since consumers do not always value them fully. $CaCO_3$ isn't recommended because the presence of CO_2 triggers foaming and an uncontrolled favourable pH, but $Ca(OH)_2$ may result in some precipitation. Despite the fact that they have not yet caused any tactile distaste, this methodology may be prohibited by laws and could lead to chemophobic reactions in individuals who object to any substance introduction in everyday things. (Vera *et al.* 2003).

Utilization of ordinary ED stacks made by substituting cationic and anionic films was proposed to lessen the causticity of a few natural products, like grape, orange, lemon, and pineapple and ume (salted Japanese apricot) (Ono *et al.* 1992) juices with practically no compound expansion. It was discovered that a more thorough clarification of organic product juices using ultrafiltration might reduce film fouling.

Beetroot molasses is the elementary result as a by-product in the sugar-producing process. In spite of the typical half sugar content (w/w) of the substrate, further sugar recuperation is hampered by the presence of contamination to inorganic salts from suspended materials and various other substances. A lot of emphasis has been placed on soluble base metal cations, specifically Na^+, K^+, and Ca^{2+}; furthermore, Mg^{2+} ions are viewed as melassigenic particles (Elmidaoui *et al.* 2004). To dispose of multivalent cations, molasses is by and large incorporated with lime and warmed, causing precipitation of $CaSO_4$ and $MgSO_4$, which are then isolated by centrifugation. A few elective cycles, for example, IER, engineered adsorbents, coagulants, membrane filtration, and ED, have been proposed up to this point. For instance, sugarcane juice may be clarified through filtration (Thampy *et al.* 1999) or MF (Pinacci *et al.* 2004) to limit fouling peculiarities. It is then concentrated to around 30°Brix, desalted by 50–80% using traditional ED, further focused, and then solidified, resulting in gems of uniform size and molasses brown in variety, but aesthetically pleasing (Thampy *et al.* 1999). The principal bottlenecks of ED application in the sugar business are both the short layer time, particularly for anion exchange membranes, and the high consistency of stick or beet-sugar syrups, the most extreme working temperature for electro-membranes being by and large under 40°C.

2.5.2 Electrodialysis in Wastewater Treatment

ED was an early method of water desalination and it is still widely recommended in this field for its high-water recuperation properties, long lifetime, and adequate power utilization. Today, on account of mechanical advancement in ED processes and the rise of ion exchange membranes (IEMs), ED has numerous applications in the food business. The growth of the industry has also led to several problems, such as the lifetime impediment of IEMs due to various maturation peculiarities (caused by natural and extra mineral combinations). The ongoing business IEMs show amazing execution in ED processes; however, natural foulants like surfactants, proteins, polyphenols, and other regular natural matter can stick on a surface (particularly while utilizing anion exchange membranes: AEMs), framing a colloid layer that can invade the membrane framework, which prompts expansion in electrical obstructions, bringing about higher energy consumption, lower water recovery, and loss of membrane perm-selectivity and momentum.

ED is an electro-membrane technique that aims at the modification of the ionic composition of liquids by using ion permeability. Traditional and capricious ED designs has been tried to treat a few wastes or spent fluid arrangements, including effluents from different modern cycles, civil wastewater or saltwater management plants, and mammal ranches. Characteristics like selectivity, high detachment productivity, and compound-free treatment makes ED strategies sufficient for desalination and different medicines with critical natural advantages. ED advancements can be utilized in activities of fixation, weakening, desalination, recovery, and valorization techniques to recover wastewater and to recuperate water and additionally different items, such as weighty metal particles, salts, acids/bases, supplements, and organics, or electrical energy. Extreme exploration movements have been coordinated in the direction of creating upgraded or new frameworks, showing the zero or insignificant fluid release approaches can be technologically and financially reasonable. Regardless of some genuine plants have been introduced, late advancements are resulting in opening novel courses for a huge scope utilization of ED methods in plenty of wastewater treatment processes (Luigi *et al.* 2020).

2.6 Applications of Electrodialysis with Bipolar Membranes in Food Processing

Three sorts of cell arrangements of bipolar membrane ED and two sorts of cell arrangements of ED were applied for extracting gluconic acid from sodium gluconate. For example, ongoing variety, change rate, effectiveness of current, and utilization of energy were looked at (Lei *et al.* 2020). Consistently, sweet whey is produced in enormous amounts, and valorization of it is frequently joined by using a preliminary skimming step following a centrifugation process. However, a few leftover lipids might be available in the wake of the skimming cycle and result in mechanical troubles. ED with bipolar film might be utilized to eliminate these leftover lipids, at the same time diminishing ionic strength and pH and permitting the development of lipoprotein edifices that additionally have to be recuperated (Faucher *et al.* 2021).

Caseins are delivered by a number of strategies like ultracentrifugation, microfiltration, and rennet coagulation, and the most commonly utilized technique is precipitation to reach its isoelectric point. Another technique coupling ED with bipolar films and an ultrafiltration unit (EDBM-UF) has recently been created for casein and caseinate production as an option in contrast to synthetic fermentation. In the innovate mixture, milk passes through an ultrafiltration film, which permits the holding of milk proteins—casein. The UF penetrates, made out of lactose and minerals, flow into the EDBM cell, and when an electrical flow is applied, UF pervade is electro-acidified by the H+ ions created by the bipolar membrane. Moreover, the pervade is demineralized because of the relocation of cations through the cation-trade film. The fermented and demineralized saturate is utilized to diminish milk pH and encourage casein production (Deschenes *et al.* 2022).

Table salt preparation from seawater by ED utilization can reach sodium chloride production up to 200 kg/m^3, preceding dissipation and salt crystallization, and has a specific business significance, particularly in Japan and Kuwait (Strathmann 1991). The resulting outcome of the innovation has been minimal expense, particularly of conductive films with a preferred penetrability of mono-valent particles. This permits chloride particles to be accumulated in the concentrated stream, while Ca^{2+}, Mg^{2+} particles, and sulphates were completely dismissed in the weakening stream.

An area where the utilization of ED is potentially intriguing is that of the fermentation industry, especially when the primary result of microbial digestion is an electrolyte. This might have an inhibitory impact on cell development or potentially metabolite creation in either its free or separated structure. On the other hand, it might be broken down in a media-rich of debasements that are to be taken out through various and costly decontamination steps. With membrane recycle bioreactors (MBRs) (Parente *et al.* 1995; Enzminger and Asenjo 1986), MF modules permits evacuation of the repressing metabolites, thus helping to boosting cell thickness in the bioreactor, as well as bioproduct development rate. Further ED treatment of MF pervades leads to two streams, a weakened one to be reused once more in the bioreactor, and a concentrated one to be valuably refined.

Although the ED business has had a consistent development pace of around 15% in the last 15 years (Srikanth 2004) and ED has various possibilities, its applications are still limited in the food business. Why is it so rarely used? According to custom, the specialised development of the food industry has often progressed gradually, with a 20–30 year delay compared to that of the chemical and pharmaceutical industries (Cantarelli 1987). For example, it is significant that falling film evaporators began to be utilized in the Italian citrus industry by the mid-1970s even though they had been mechanically fabricated in 1947 by the Mojonnier Brothers Company for the Florida Citrus Canners Cooperative (Varsel 1980) as a development of the strategies for fixation under vacuum created during World War II to focus on very thermosensitive materials, like penicillin. To check the food industry of the world, it is important to fall back on proper increasing practices in pilot or modern plant scale to evaluate unequivocally the film interaction execution and real capacity, as well as its prudent plausibility. There are various issues that have definitely restricted development in ED membrane deals, such as membrane fouling issues, plan contemplations, cleanability, venture furthermore, membrane substitution costs, and contending innovations.

Membrane fouling and scaling are brought about by dissolvable and insoluble impurities present in the feed, for example, insoluble salts, natural matter, colloidal substances, and microorganisms. Though the anionic membranes appear to be more likely to be fouled by natural matter, the cationic ones will generally be scaled by inorganic matter. To limit such issues, as well as stack stopping, feed pre-treatment through MF, UF, NF, or IER might be helpful. Other approaches, like changing membrane properties (Grebenyuk et al. 1998) or the use of beat electric fields (Lee et al. 2003), appear to be convenient to ease anionic membrane fouling. In the majority of cases, cationic membrane scaling was viewed as practically reversible and taken care of by pH change, EDT A, or citrus extract expansion. Subsequently, there is still no obvious technique to direct fouling and scaling in ED handling, and there is a need for extra research on this issue so that work can be advanced considering scaling and fouling.

2.7 Applications of Electrodialysis with Filtration Membranes in Food Processing

Another innovation in ED with an ultrafiltration membrane technique (EDUF) or conventional ED with a filtration membrane (EDFM) was created by Bazinet and Firdaous (2013) and patented. This process has benefits of size prohibition capabilities of permeable membranes with the charge selectivity of ED. In this process, a regular ED cell is utilized, whose few IE membranes are supplanted by filtration membranes (microfiltration, nanofiltration, or ultrafiltration); mixtures of molecular weights higher than the film size can be

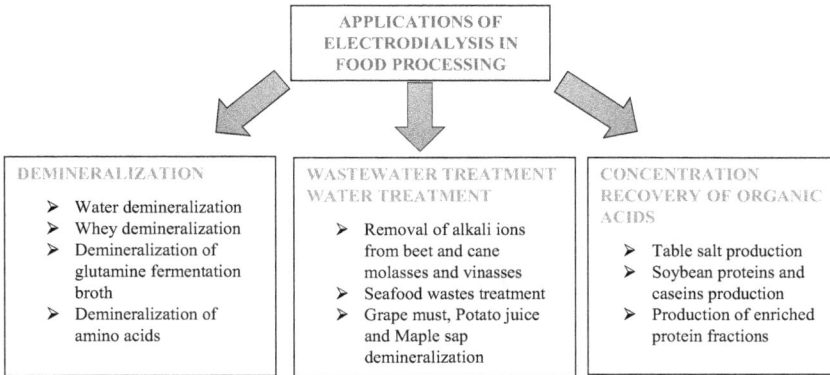

APPLICATIONS OF
ELECTRODIALYSIS IN
FOOD PROCESSING

DEMINERALIZATION

➤ Water demineralization
➤ Whey demineralization
➤ Demineralization of
   glutamine fermentation
   broth
➤ Demineralization of
   amino acids

WASTEWATER TREATMENT
WATER TREATMENT

➤ Removal of alkali ions
   from beet and cane
   molasses and vinasses
➤ Seafood wastes treatment
➤ Grape must, Potato juice
   and Maple sap
   demineralization

CONCENTRATION
RECOVERY OF ORGANIC
ACIDS

➤ Table salt production
➤ Soybean proteins and
   caseins production
➤ Production of enriched
   protein fractions

Figure 2.4 Electrodialysis in food processing.

isolated by expanding the field of ED utilization to organic charged particles.
Without strain application in the ED cell, only charged atoms move under the
impact of the electric field, and the unaligned particles remain hypothetically
in the feed arrangement and do not pass the filtration membrane (Bazinet and
Firdaous 2013). The various applications of ED are presented in Figure 2.4.

2.8 Conclusion

Electrodialysis is an electro-membrane technology that utilizes ion-
permeable membranes along with direct electrical potential to modify the
ionic composition of liquids. ED has evolved over the past century from
a lab experiment to a powerful device with a variety of industrial uses.
Particularly noteworthy is the use of various ED approaches for difficult
separations observed in various food and beverage businesses. Because of
their selectivity, ion membranes permit the transport of only cations (cation-
exchange membranes) or anions (anion-change films) and in this way can
be utilized to concentrate, eliminate, or separate electrolytes. In addition to
being a technique for desalination, ED in all of its forms has emerged as a
practical and affordable solution to separation challenges in the dairy, juice,
beverage, and wine sectors.

References

Alvarez, F., Alvarez, R., Coca, J., Sandeaux, J., Sandeaux, R., & Gavach, C.
(1997). Salicylic acid production by electrodialysis with bipolar membranes. *Journal of Membrane Science*, **123**, 61–69.

Audinos, R. (1992). Liquid waste concentration by electrodialysis. In *Separation and purification technology*, pp. 229–301, Marcel Dekker, New York.

Batchelder, B. T. (1987). Electrodialysis applications in whey processing. *Bulletin-Federation Internationale de Laiterie (Belgium)*. *International Dairy Federation*, no. 212, ISSN: 0250–5118.

Bazinet, L. F. L. L., & Firdaous, L. (2013). Separation of bioactive peptides by membrane processes: Technologies and devices. *Recent Patents on Biotechnology*, **7**, 9–27.

Bazinet, L. F. L. L., Lamarche, F., & Ippersiel, D. (1998). Bipolar-membrane electrodialysis: Applications of electrodialysis in the food industry. *Trends in Food Science & Technology*, **9**, 107–113.

Boyaval, P., Seta, J., & Gavach, C. (1993). Concentrated propionic acid production by electrodialysis. *Enzyme and Microbial Technology*, **15**, 683–686.

Cantarelli, C. (1987). Ricerca e formazione nel campo delle biotecnologie alimentari. *Industrie Alimentari*, **26**, 333–349.

Cassel, G. E., & Kempe, D. (1894). Verfahren: Melasse, Sirop und andere Zuckerlösungen elektrolytisch zu reinigen, German Patent No 78972, 10.

Chen, G. Q., Eschbach, F. I., Weeks, M., Gras, S. L., & Kentish, S. E. (2016). Removal of lactic acid from acid whey using electrodialysis. *Separation and Purification Technology*, **158**, 230–237.

Deschênes Gagnon, R., Bazinet, L., & Mikhaylin, S. (2022). Functional properties of casein and caseinate produced by electrodialysis with bipolar membrane coupled to an ultrafiltration module. *Membranes*, **12**, 270.

Donnan, F. G. (1995). Theory of membrane equilibria and membrane potentials in the presence of non-dialysing electrolytes. A contribution to physical-chemical physiology. *Journal of Membrane Science*, **100**, 45–55.

Durham, R. J. (2009). Modern approaches to lactose production. *Dairy-Derived Ingredients*, 103–144.

Elmidaoui, A., Chay, L., Tahaikt, M., Sahli, M. M., Taky, M., Tiyal, F., Khalidi, A., & Hafidi, M. R. A. (2004). Demineralisation of beet sugar syrup, juice and molasses using an electrodialysis pilot plant to reduce melassigenic ions. *Desalination*, **165**, 435.

Enzminger, J. D., & Asenjo, J. A. (1986). Use of cell recycle in the aerobic fermentative production of citric acid by yeast. *Biotechnology Letters*, **81**, 7–12.

Faucher, M., Perreault, V., Gaaloul, S., & Bazinet, L. (2021). Defatting of sweet whey by electrodialysis with bipolar membranes: Effect of protein concentration factor. *Separation and Purification Technology*, **251**, 117248.

Fidaleo, M., & Moresi, M. (2006). Assessment of the main engineering parameters controlling the electrodialytic recovery of sodium propionate from aqueous solutions. *Journal of Food Engineering*, **76**, 218–231.

Grebenyuk, V.D., & Grebenyuk, O.V. (2002) Electrodialysis: From an Idea to Realization. *Russian Journal of Electrochemistry*, **38**, 806–809.

Grebenyuk, V. D., Chebotareva, R. D., Peters, S., & Linkov, V. (1998). Surface modification of anion-exchange electrodialysis membranes to enhance anti-fouling characteristics. *Desalination*, **115**, 313–329.

Hestekin, J., Ho, T., & Potts, T. (2010). Electrodialysis in the food industry. *Membrane Technology: Volume 3: Membranes for Food Applications*, **3**, 75–104.

Huang, C., & Klein Xu, T. (2006). Electrodialysis with bipolar membranes for sustainable development. *Environmental Science & Technology*, **40**, 5233–5243.

Hughes, M., Raubenheimer, A. E., & Viljoen, A. J. (1992). Electrodialysis reversal at Tutuka power station, RSA-seven years' design and operating experience. *Water Science and Technology*, **25**, 277–289.

Juda, W., & McRae, W. A. (1950). Coherent ion-exchange gels and membranes. *Journal of the American Chemical Society*, **72**, 1044–1044.

Klein, E., Ward, R. A., & Lacey, R. E. (1987). Membrane processes—dialysis and electrodialysis. In R. W. Rousseau (ed.), *Handbook of separation process technology*, pp. 954–981, John Wiley & Sons, Wiley-Interscience Publication, New York.

Koros, W. J., Ma, Y. H., & Shimidzu, T. (1996). Terminology for membranes and membrane processes (IUPAC Recommendations 1996). *Pure and Applied Chemistry*, **68**, 1479–1489.

Kressman, T. R. E. (1950). Ion exchange resin membranes and resin-impregnated filter paper. *Nature*, **165**, 568–568.

Krol, J. J. (1999). Monopolar and bipolar ion exchange membranes: Mass transport limitations. 0132–0132.

Lacey, R. E., & Loeb, S. (1972). *Industrial processing with membranes*, pp. 21–106, Wiley-Interscience Publication, John Wiley & Sons, New York.

Lee, H. J., Oh, S. J., & Moon, S. H. (2003). Recovery of ammonium sulfate from fermentation waste by electrodialysis. *Water Research*, **37**, 1091–1099.

Lei, C., Li, Z., Gao, Q., Fu, R., Wang, W., Li, Q., & Liu, Z. (2020). Comparative study on the production of gluconic acid by electrodialysis and bipolar membrane electrodialysis: Effects of cell configurations. *Journal of Membrane Science*, **608**, 118192.

Maigrot, E., & Sabates, J. (1890) Apparat zur Laiuterung von Zuckersaiften mittels Elektrizitiit, German Patent No. 50443.

Manegold, E., & Kalauch, C. (1939). Uber Kapillarsysteme. XXII Die Wirksamkeit Verschiedener Reinigungsmethoden (Filtration, Dialyse, Electrolyse und Ihre Kombinationen). *Kolloid Z.*, **86**, 93.

Mani, K. N. (1991). Electrodialysis water splitting technology. *Journal of Membrane Science*, **58**, 117–138.

Meyer, K. H., & Straus, W. (1940). La perméabilité des membranes VI. Sur le passage du courant electrique a travers des membranes sélectives. *Helvetica Chimica Acta*, **23**, 795–800.

Mondor, M., Ippersiel, D., & Lamarche, F. (2012). Electrodialysis in food processing. In *Green technologies in food production and processing*, pp. 295–326, Springer, Boston, MA.

Mulder, M., & Mulder, J. (1996). *Basic principles of membrane technology*, Springer Science & Business Media, New York.

Nagasubramanian, K., Chlanda, F. P., & Liu, K. (1977). Use of bipolar membranes for generation of acid and base — an engineering and economic analysis. *Journal of Membrane Science*, **2**, 109–124.

Ono, T., Teramoto, T., & Sawada, M. (1992). *Electrodialysis of fruit juices for reduction of acidity.*

Ostwald, W. (1890). Elektrische eigenschaften halbdurchlässiger scheidewände. *Zeitschrift für physikalische Chemie*, **6**, 71–82.

Parente, E., Ricciardi, A., Mancino, M., & Moresi, M. (1995). Repeated batch citrate production by *Yarrowia lipolytica* using yeast recycling by aseptic centrifugation. Annali di Microbiologia ed Enzimologia (Italy).

Pinacci, P., Radaelli, M., Bottino, A., & Capannelli, G. (2004). Molasses purification by integrated membrane processes. *Filtration* (Coalville, United Kingdom), **4**, 119–122.

Ronald, W. (1987). *Handbook of separation processes technology*, R. W. Roussean, ed., p. 1024, Wiley-Interscience Publication, New York.

Schaffer, L. H., & Mintz, M. S. (1966). Electrodialysis. In K. S. Spiegler (ed.), *Principles of desalination*, pp. 3–20, Academic Press, New York.

Schollmeyer, G. (1902). Reinigung von Znckersaften durch Elektrodialyse und mit Ozon. German Patent No. 136670.

Shaposhnik, V. A., & Kesore, K. (1997). An early history of electrodialysis with permselective membranes. *Journal of Membrane Science*, **136**, 35–39.

Srikanth, G. (2004). Membrane separation processes—Technology and business opportunities. In *News and views, technology information, forecasting & assessment council.* www.tifac.org.in/news/memb.htm.

Staff, A. W. W. A. (1996). *Electrodialysis and electrodialysis reversal (M38)*, American Water Works Association, United States.

Strathmann, H. (1991). Electrodialysis in membrane separation systems. In R. W. Baker, E. L. Cussler, W. Eykamp, W. J. Koros, R. L. Riley, & H. Strathmann (eds.), *Membrane separation systems*, pp. 396–420, Noyes Data Corp., Park Ridge, NJ.

Strathmann, H. (1995). Chapter 6: Electrodialysis and related processes. In Richard D. Noble & S. Alexander Stern (eds.), *Membrane science and technology*, Elsevier, United States, **2**, pp. 213–281.

Strathmann, H. (2004). Assessment of electrodialysis water desalination process costs. In *Proceedings of the international conference on desalination costing, Limassol, Cyprus*, pp. 32–54.

Strathmann, H. (2005). Membranes and membrane separation processes. In Ullmann's Encyclopedia of Industrial Chemistry (ed.). https://doi.org/10.1002/14356007.a16_187.pub2.

ELECTRODIALYSIS

Teorell, T. (1953). Transport processes in ionic membranes. *Progress in Biophysics and Biophysical Chemistry*, **3**, 305.

Thampy, S. K., Narayanan, P. K., Trivedi, G. S., Gohil, D. K., & Indushekhar, V. K. (1999). Demineralization of sugar cane juice by electrodialysis. *International Sugar Journal*, **101**, 365–366.

Valero, F., Barceló, A., & Arbós, R. (2011). Electrodialysis technology-theory and applications. *Desalination, Trends and Technologies*, **28**, 3–20.

VanNispen, J. G. M. (1991) US patent 5 002 881.

Varsel, C. (1980). Citrus processing as related to quality and nutrition. In S. Nagy & J. A. Attaway (eds.), *Citrus nutrition and quality*, pp. 225–271, ACS Symposium Series no. 143, American Chemical Society, Washington, DC.

Vera, E., Ruales, J., Dornier, M., Sandeaux, J., Sandeaux, R., & Pourcelly, G. (2003). Deacidification of clarified passion fruit juice using different configurations of electrodialysis. *Journal of Chemical Technology & Biotechnology: International Research in Process, Environmental & Clean Technology*, **78**, 918–925.

Wang, Y., Jiang, C., Bazinet, L., & Xu, T. (2019). Electrodialysis-based separation technologies in the food industry. In *Separation of functional molecules in food by membrane technology*, pp. 349–381, Academic Press, United States.

Xu, T. (2005). Ion exchange membranes: State of their development and perspective. *Journal of Membrane Science*, **263**, 1–29.

Young, P. (1974). An introduction to electrodialysis. *Journal of Society of Dairy Technology*, **27**, 141–150.

Chapter 3

Refractance Window Technology as a Promising Drying Technique in the Food Industry

Kirti Jalgaonkar, Manoj Kumar Mahawar, and Srinivas Girijal

3.1 Introduction

Drying is considered an imperative post-harvest unit operation concerned with removing moisture from products by the application of natural/artificial heat. The resultant dried product is typically obtained as intermediate moisture foods, sheets, flakes, film, powder, or granules. Basically, drying is a thermo-physical and physico-chemical process and therefore an energy-intensive technique (Raghavi et al. 2018; Verma et al. 2020). For decades, its superlative applications were implemented in the food industry to curtail the growth of microorganisms, minimizing undesirable chemical reactions, providing safer storage conditions, and reducing the expenses of transportation. Despite such intriguing features, several detrimental changes in the physical characteristics (texture, color, shrinkage, deformation, crack

DOI: 10.1201/9781003217138-3

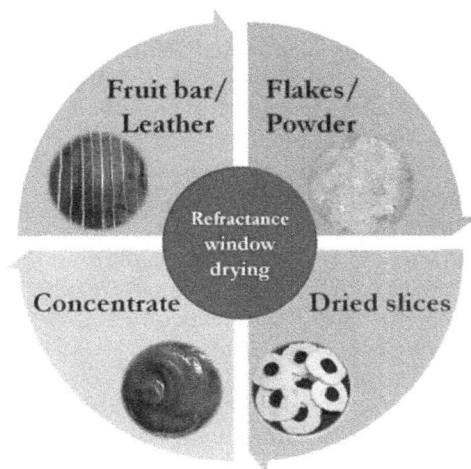

Figure 3.1 Different products obtained from RW drying.

development) may occur in the final product. Such changes will definitely affect the sensorial, nutritional, and chemical properties (total polyphenol, antioxidant activity, etc.) and ultimately consumer acceptability (Calín-Sánchez et al. 2020). Substantial degradation and loss of different bioactive compounds from the food products result in the slashing of health-beneficial characteristics. The application and adoption of such preservation techniques that minimally alter the intrinsic quality with easier scalability potential is an urgent requirement in the food industry.

Refractance window (RW) drying is a relatively novel technique with the working principle of high temperature/short-time process and is mostly used for the preparation of intermediate-moisture foods with enhanced nutrient retention. This method has become the most-used drying method in the food industries for moisture removal from heat-sensitive fruit and vegetable products. The end products can be in the form of fruit bar/leather, slices, flakes, powders, and concentrates, as shown in Figure 3.1 (Mahanti et al. 2021). The RW setup is comparatively easier to install and requires inexpensive equipment with lower energy consumption. In comparison to freeze dryers, the energy consumption is about one-third to one-half lower for drying a similar quantity of moist products (Nindo and Tang 2007).

Depending upon the requirements, the RW drying system can be customized in batch or continuous form. Mylar film constitutes an important part of the RW drying method which allows infrared radiation to pass and is also helpful in creating a window for drying. The Mylar film used for RW drying is a food-grade polyester film. Hot water (95–98°C) at atmospheric pressure is used as a heating medium, enabling heat transfer for moisture evaporation from the product. A heat-stable Mylar film is placed above the hot water surface, and the transfer of thermal energy to the material takes place in the form of a thin layer over the Mylar sheet. The sheet functions to restrict the evaporation of water and its associated heat loss, during which only conduction occurs (Magoon 1986).

While operating, the wet product is spread on the Mylar film, preferably as slices or a thin uniform layer of pulp, allowing the infrared energy from the hot water to pass through. At this juncture, all three modes of heat transfer occur, which leads to rapid evaporation of water and faster drying. As complete drying is achieved, the infrared window closes effectively, limiting further heat transfer to the product. The thermal energy of hot water is transmitted to the wet product by means of conduction and radiation through the plastic film. Temperature rise beyond 70°C is restricted owing to the lower thermal conductivity of the Mylar sheet and the resultant concurrent cooling of the product because of evaporation.

In batch-type RW systems, usually a thermostatic water bath is utilized and the Mylar sheet is fitted over a customized perforated tray, preferably of stainless steel. However, in a continuous RW system, the Mylar sheet is in the form of a moving belt. Movement of the belt can be achieved with the help of pulleys driven by a motor. The belt speed can be controlled using a set of a gearbox and speed regulator. For uniform drying, the belt speed has to be kept to a minimum (0.6 to 3 m/min) (Raghavi et al. 2018) so that the contact time of the wet product with hot water is increased, thereby enhancing the rate of moisture removal, resulting in very rapid drying. A provision of doctor blades that span the full width of the belt should be given at the endpoint to scrape off the dry product. If the recovered product is in the form of dried flakes or powder, it has to be immediately packed into moisture-proof packets to avoid moisture absorption from the ambient atmosphere. Intermediate-moisture foods like fruit bars and fruit leather can also be prepared by modifying the pulp constituents. After the required moisture removal, the semi-dried layer can be cut into desired shapes for further packaging and storage.

The intriguing features of RW drying include drying at a relatively lower temperature, simple design, energy efficiency, low cost, faster drying, and maximum retention of desired bioactive compounds responsible for the sensorial properties of the final product. The retention of flavor; color; and aromatic, nutritional, and bioactive compounds is reported to be higher in RW drying, and the product quality is also equivalent to or sometimes superior to that of freeze drying. RW drying was developed and patented by MCD Technologies, Tacoma and Mt. Capra, Washington (Magoon 1986). The first literature in the form of a published paper on RW drying of strawberries and carrots was reported by Abonyi et al. (2002).

3.2 Novelty

Refractance window drying is regarded as a fourth-generation drying process and has diversified applications for different type of products. It is also called conductive hydro-drying (Ortiz-Jerez and Ochoa-Martinez 2015) or cast tape drying (Durigon et al. 2016). The heated air around the periphery of the dryer assists in moisture removal from the product. Reports suggest that the overall requirement is lower to the tune of 50–70% (investment) and 50% (energy) when compared with the same quantity of lyophilized material (Nindo and Tang 2007).

The application of hot water below its boiling point as the heating medium is an addition to the novelty of this drying method. The evaporative cooling which is the possible outcome of the development of an air gap and associated effects at the product surface assist in maintaining the temperature between 70–85°C. At this temperature, a dried product with superior quality characteristics can be preserved, enabling its extended applications for high-value products. The quality characteristics and heat-sensitive nutritive composition can be intuitively restored in the final product. Several researchers have compared the effect of refractive drying with existing drying methods and reported the advantages with respect to superior preservation of heat-sensitive compounds, quality attributes, energy efficacy, and lower operating expenses.

Nindo and Tang (2007) presented their views on the working mechanism of RW drying. They stated that without any product over the Mylar film, the refractive index between water and ambient air did not match. This results in reflecting radiant energy at the plastic–water interface back towards the water and comparatively less transfer through the sheet to the atmosphere (window closed). As soon as any moist product is placed over the film, the mismatch of refractive index between water and food drops significantly (i.e., water–plastic–food system). At this point, the converging of refractive indices results in decreased reflection, therefore increasing the transmitting rate of radiant energy into the food (Yoha et al. 2019). With the progression in the drying phenomenon, the product moisture decreases and subsequently leads to an increased mismatch in the refractive index. As a result, a substantial amount of radiant energy is transmitted back into the water, ultimately stopping drying. Ortiz-Jerez and Ochoa-Martinez (2015) also corroborated that the air gap between the product and sheet caused a reduced temperature of the product. This was a consequence of shrinkage in the product and the convective cooling caused by the flow of ambient air over the drying product. The energy requirement is predominantly for the evaporation of water from the air–product interface in RW drying (Nindo and Tang 2007). The factors affecting RW drying are shown in Figure 3.2.

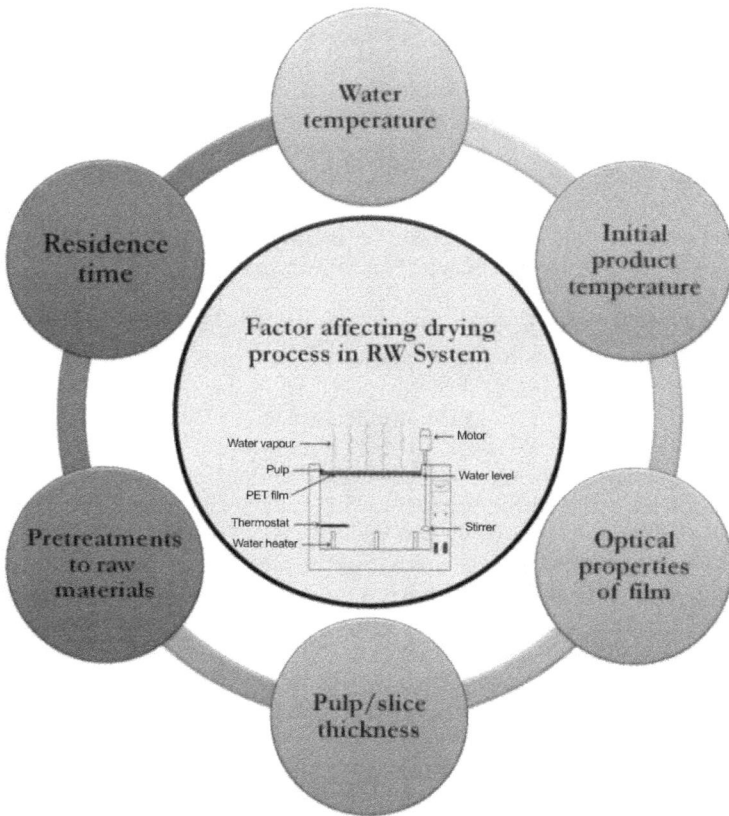

Figure 3.2 Factors affecting RW drying.

3.3 Principles and Theory

The unique feature of refractance window drying lies in the fact that it employs circulating hot water (95–97°C) at ambient conditions which transfer thermal energy to the product subjected to drying through the Mylar film. During this process, the product temperature doesn't exceed 70°C. Products like pulp, purees, juice, and suspensions are spread on the Mylar sheet, having contact with hot water at the lower side. Since water is utilized as the heating medium, it can be easily recycled for reuse, thereby improving the overall thermal efficiency of the process.

Basically, all three principal heat transfer modes are actively responsible for enhanced and rapid moisture removal. Water heating is accomplished within an insulated tank (glass wool), and then hot water is circulated to the thin plastic conveyor, which reaches the temperature of hot water quickly and facilitates rapid drying. As opined by the inventors of RW drying, the contact of plastic film with hot water on one side and the presence of a wet product on the other results in increased infrared transmission.

When a wet product is spread on the Mylar film, refraction at the plastic–product interface is restricted, thereby enabling radiant thermal energy to pass into the product. The thickness and initial moisture of the product affect the heat transfer (Ratti and Mujumdar 1995). Heat transfer during the process is predominantly by evaporative cooling and convection of the product. This evaporation is very intense and is responsible for ample energy consumption during drying. In the later stages, when the product is nearing dry, the conduction mode becomes active, and gradually the heat transfer rate to the product slows down with further drying. After complete drying, the cooling section at the discharge end will facilitate reducing the product temperature for easy removal and further storage.

3.4 Equipment Description

Jalgaonkar et al. (2020) developed a batch-type RW setup and evaluated its performance for sapota pulp drying. The setup has the capacity of holding 60 liters of water in a thermostatic water bath. A stainless steel frame was fitted with the polyethylene terephthalate (PET) sheet (80 μm thickness) for the drying of the pulp, as shown in Figure 3.3. A diagram of a continuous RW system is also shown in Figure 3.4.

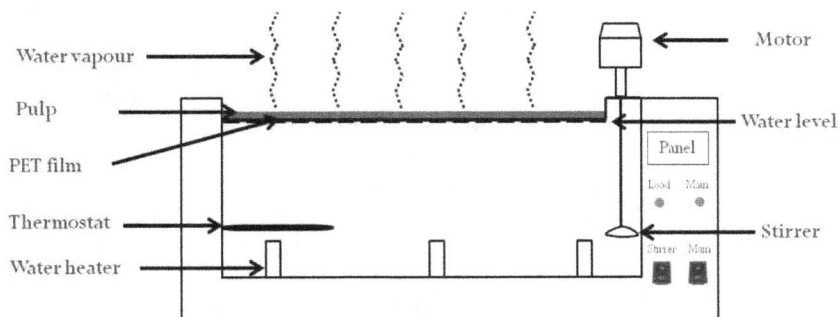

Figure 3.3 Schematic sketch of batch type RW system.

Source: Jalgaonkar et al. (2020)

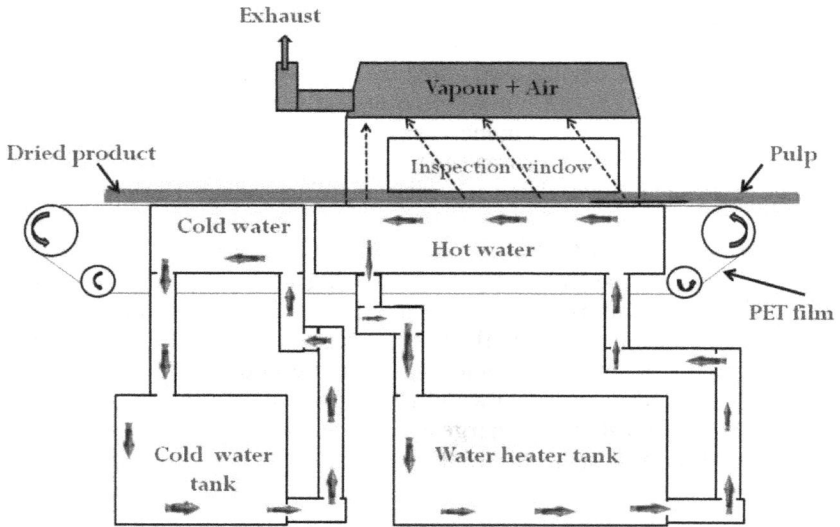

Figure 3.4 Schematic working setup of a continuous RW system.

3.5 Impact on Various Food Properties

Bioactive compounds comprise phytochemicals, phenolic compounds, flavors, pigments, essential oils, and nutraceuticals (Zia et al. 2020). Limited studies are there in the literature iterating on the effect of RW drying on such bioactive compounds. Density and porosity are the properties of the dried product which are most affected by RW drying. The bulk density values are generally higher when compared to those of products obtained using freeze and spray drying. The characteristic features of the dried product include lower porosity, higher bulk density, less oxidative degradation, and lower surface area, making the product suitable for storage and transportation.

The selection of an appropriate drying method is imperative to preserve the natural color and appearance of the product. Features like lower drying temperature and drying time of RW drying contribute to maximal color retention of the dried product. Therefore, the applicability and suitability of RW drying are highest for food products where color retention is one of the primary requirements.

Morphological studies have shown the presence of smooth surface in RW dried products such as aloe vera powder (Minjares-Fuentes et al. 2017); mango powder (Caparino et al. 2012; Shende and Datta 2020); haskap berry powder (Celli et al. 2016); blueberry, tart cherry, strawberry, and cranberry powder (Nemzer et al. 2018); apple slices (Hernández et al. 2019; Rajoriya et al. 2019); and tomato powder (Qiu et al. 2019). Moreover, the products are less prone to oxidation, and the smooth surface was favorable for free flow. Ochoa-Martinez et al. (2012) reported higher effective diffusivity in RW drying than that in tray drying irrespective of the thickness of mango slices. RW-dried products have superior textural and functional properties when compared with products obtained using other conventional drying techniques. Thorough studies are required to understand the microbial load of RW-dried products. The properties affected by RW drying are discussed in the following sections:

3.5.1 Ascorbic Acid Retention

Vitamin C is the most sensitive nutrient to excessive heat and is generally degraded while drying. Its maximum retention in the dried product represents a good indication of optimized drying and better-quality product. Abony et al. (2002) reported the versatility of RW drying in retaining ascorbic acid (AA) when compared with freeze drying of strawberry puree. RW drying provided the best protection for ascorbic acid in the pomegranate pestil against oxidation reaction (Tontul and Topuz 2017). Tontul et al. (2018b) evaluated the dried product of cornelian cherry pulp obtained using convective drying (50, 60, and 70°C) and RW drying (90, 95, and 98°C). The results showed that vitamin C retention was higher (0.9 g/kg dm) of RW dried product as compared with hot air dried product (0.8 g/kg dm). Jalgaonkar et al. (2020) found that ascorbic acid content in the final product was positively correlated with water temperature from 84.3 to 91°C, and further increasing the temperature beyond 91°C resulted in a substantial decrease. Rajoriya et al. (2019) reported that drying of apple slices using RW resulted in the highest (80%) retention of vitamin C, higher than that of hot air drying (69%) and equivalent to freeze drying (83%). Shende and Datta (2020) reported the loss of ascorbic acid in mango pulp was lowest in RW drying (33.68%), followed by tray drying (65.26%) and oven drying (70.53%). Leiton-Ramirez et al. (2020) dried guava pulp using different methods and reported substantial loss of ascorbic acid (92.30%) in tray drying, followed by RW (69.57%) and freeze drying (62.13%). Rajoriya et al. (2021) studied the effect of banana puree thicknesses (2 and 3 mm) using RW drying (70 to 90°C) and reported the retention of ascorbic acid in both samples.

3.5.2 Total Phenolic Content Retention

Different drying parameters have a profound influence on phenolic compounds. Particularly in RW drying, processing conditions like light exposure, oxygen availability, and water temperature affect the preservation of phenols in the product (Papoutsis et al. 2018). Shende and Datta (2020) emphasized that water temperature and thickness of mango pulp in a RW drying system significantly affected the phenolic content of mango leather. The variation in temperature of water from 85, 90, and 95°C resulted in the respective increase in total phenolic content (TPC) from 4.9, 6.3, and 7.5 mg GAE/g. Baeghbali et al. (2019) revealed that the percentage loss of TPC in apple slices was lowest in RW drying (7.01%) compared to ultrasound and infrared-assisted conductive RW drying (8.32%). Baeghbali et al. (2016) reported that drying of pomegranate juice showed the minimum loss in TPC using RW drying (7.94%) compared to the spray drying method (11.79%). Nayak et al. (2011) reported that RW-dried colored potato flakes had significantly higher TPC (4680 mg GAE/g DW) in comparison with drum drying (4021 mg GAE/g DW) and freeze drying (3950 mg GAE/g DW).

3.5.3 Total Anthocyanin Content Retention

Anthocyanins are comparatively more heat sensitive than phenolic compounds. Tontul and Topuz (2017) emphasized that a temperature increment from 50 to 70°C in hot air drying resulted in the reduction of total anthocyanin content (TAC) of 15.54 to 10.14 mg of cyanidin-3-glucoside (C3G) equivalent per 100 g of dry matter. On the contrary, in RW drying with variation in water temperature from 90, 95, and 98°C, TAC improved from 11.72, 12.11, and 13.03 mg of C3G per 100 g of dry matter. Baeghbali et al. (2016) compared the TAC content of pomegranate juice and observed a lower reduction in the RW dried sample (14.99%) than freeze-dried (28.83%) and spray-dried (39.04%) juice samples. The results substantiated that RW dried products had significant retention and concentration of anthocyanin.

3.5.4 Carotenoid Retention

Abonyi et al. (2002) reported a non-significant difference in carotenoid content between RW and lyophilized carrot puree. The α-carotene, β-carotene, and total carotene losses for RW dried samples were 7.4%, 9.9%, and 8.7%, while the corresponding values for lyophilized samples were 2.4%, 5.4%, and 4.0%. Leiton-Ramirez et al. (2020) reported a decrease in total carotenoids of 24.36% (RW drying), 28.95% (freeze drying), and 23.20% (tray drying). da Costa et al. (2019) reported that RW drying conditions (70 ± 1°C and 0.5 mm pulp thickness) resulted in the reduction of 7.9, 8.9, and 11.5% of total carotenoids in edible film prepared using peach palm and tucupi when compared to their blends. Overall, it can be inferred that carotenoids and capsaicinoids are prone to degradation in RW drying.

3.5.5 Total Antioxidant Activity Retention

Rajoria et al. (2021) reported the highest retention of total antioxidant activity (TAA) (74–80%) in dried banana puree samples using RW drying and also reported that an increase in puree thickness had a negative impact on TAA. Baeghbali et al. (2019) reported about 6.81% reduction of TAA in RW dried and 28.44% reduction in cabinet dried apple slices. Baeghbali et al. (2016) found TAA of RW dried pomegranate juice on a par with freeze-dried and appreciably greater than spray-dried samples. Drying of potatoes (purple color) puree using RW, freeze, and drum drying showed a non-significant effect on TAA as compared with the control sample (Nayak et al. 2011). It can therefore be concluded that lower drying time at higher temperature results in increased retention of antioxidants in RW drying systems.

3.5.6 Organoleptic Qualities

Several researchers have highlighted the importance of RW drying in preserving the natural color and aroma of various food materials. Nindo and Tang (2007) emphasized that dried products with superior quality parameters were obtained after RW drying in terms of nutritional and organoleptic properties. Baeghbali et al. (2016) revealed that RW and freeze-dried pomegranate juice powder were more red but slightly darker than the control but superior to a spray-dried sample. Azizi et al. (2016) showed the organoleptic properties of RW dried kiwifruits were judged medium to good quality by panel experts. The inference can be made that RW drying has minimal impact on the sensorial properties of the products. A detailed comparison of color parameters of different products as influenced by drying techniques is shown in Table 3.1.

3.5.7 Texture

The final texture of the dried product is predominantly affected by the initial thickness of the material. Ochoa-Martinez et al. (2012), Jafari et al. (2016), and Tontul and Topuz (2017) highlighted that RW drying provided better texture compared to other drying methods for mango slices, kiwifruit slices, and pomegranate pestil production, respectively. Jalgaonkar et al. (2020) reported that increasing water temperature from 84.3 to 97.7°C in RW drying resulted in case hardening and increased sapota bar hardness. Shende and Datta (2020) also reported similar findings for mango leather.

Different authors have compared the drying conditions and their after-effects on the product characteristics described in Table 3.2.

Table 3.1 Comparative Description of Calorimetric Properties of Different Products Obtained Using Different Drying Techniques

Product	Fresh	RWD	DD	FD	SD	OD/TD	References
Strawberry puree							Abony et al. (2002)
L	36.1±1.0	53.8±0.3		53.8±0.5			
a	25.6±0.6	27.9±0.3		30.0±0.4			
b	19.8±0.9	16.9±0.3		18.8±0.4			
Change in color		18.08		18.27			
Strawberry puree + maltodextrin							Abony et al. (2002)
L	45.3±1.6	63.2±0.5		71.5±0.5	77.8±0.7		
a	27.0±1.7	29.3±0.6		25.6±0.8	23.9±0.6		
b	22.0±1.9	20.2±0.5		16.6±0.6	16.8±0.5		
Change in color		18.14		26.79	33.05		
Carrot puree							Abony et al. (2002)
L	54.3±0.8	72.0±0.3	67.5±0.6	77.6±0.4			
a	28.7±0.2	34.1±0.5	20.8±0.4	27.1±1.2			
b	44.0±1.0	45.1±0.8	39.4±1.7	44.1±0.4			
Change in color		18.54	16.06	23.36			

Paprika (Anaheim) powder

L	33.30±0.49	45.62±0.36	49.18±0.73	40.84±0.55	Topuz et al. (2009)
a	29.56±0.86	29.71±0.55	32.10±0.58	20.66±0.67	
b	14.47±0.58	21.88±0.55	25.67±1.05	13.40±0.73	
Change in color		14.38	19.60	11.71	

Paprika (jalapeño) powder

L	33.98±0.82	48.38±0.31	48.11±0.67	43.29±0.33	Topuz et al. (2009)
a	31.31±0.68	32.66±0.19	33.48±0.52	26.97±0.36	
b	16.17±0.99	27.71±0.43	27.23±0.92	20.00±0.41	
Change in color		18.50	18.07	10.96	

Tomato juice

L	56.6±0.5	61.0±0.6	63.2±0.5	Baeghbali et al. (2010)
a	62.5±0.4	48.6±0.1	43.9±0.8	
b	47.0±0.9	45.6±0.3	42.1±0.4	
Change in color		14.65	20.33	

(Continued)

Table 3.1 (Continued)

Product	Fresh	RWD	DD	FD	SD	OD/TD	References
Tomato ketchup							Baeghbali et al. (2010)
L	38.6±0.2	44.1±0.4		49.0±0.6			
a	49.6±0.2	54.0±0.4		53.6±0.1			
b	36.0±0.7	44.3±0.6		43.8±0.8			
Change in color		10.89		13.60			
Carrot puree							Baeghbali et al. (2010)
L	53.9±0.7	70.5±0.4		78.1±0.3			
a	27.3±0.1	33.4±0.6		26.4±0.5			
b	43.8±0.8	45.6±0.7		44.8±1.0			
Change in color		17.78		24.2			
Mango powder							Caparino et al. (2012)
L	45.12±0.02	43.95±0.02	37.73±0.01	43.74±0.06	41.59±0.07		
a	4.65±0.01	4.40±0.01	6.92±0.02	4.69±0.01	3.05±0.01		
b	41.52±0.03	41.79±0.03	36.48±0.02	40.99±0.23	36.64±0.02		
Change in color	–	1.22±0.02	9.22±0.01	1.57±0.03	6.23±0.02		

REFRACTANCE WINDOW TECHNOLOGY

Pomegranate juice				Baeghbali et al. (2016)
L	41.7±0.9	44.0±0.9	50.3±0.9	
a	17.3±0.5	24.3±0.5	13.7±0.5	
b	19.6±0.3	19.7±0.3	16.7±0.3	
Change in color				
Blueberry				Nemzer et al. (2018)
L	22.42±3.02	25.97±2.17	18.1±1.06	
a	4.83±0.45	9.36±3.21	6.22±0.87	
b	−0.35±0.11	1.77±0.95	2.17±0.49	
Change in color	6.13±2.19	9.22±1.91	2.89±0.52	
Cherry				Nemzer et al. (2018)
L	28.16±1.99	37.12±0.31	18.91±2.04	
a	22.63±1.36	34.87±0.78	12.62±1.40	
b	5.37±1.00	9.78±0.35	5.43±0.70	
Change in color	5.90±1.53	20.55±0.75	10.72±2.19	

(Continued)

Table 3.1 (Continued)

Product	Fresh	RWD	DD	FD	SD	OD/TD	References
Cranberry							Nemzer et al. (2018)
L		36.42±2.37		41.61±2.82		26.66±1.39	
a		41.49±0.99		40.28±4.05		18.51±0.58	
b		14.18±0.81		13.34±0.89		7.91±0.48	
Change in color		6.62±1.09		10.05±1.83		19.64±0.82	
Apple slices							Hernández et al. (2019)
L	72.28±2.93	80.64±2.33				80.64±1.25	
a	-9.10±2.93	-8.54±0.66				–7.85±0.34	
b	18.08±0.80	23.91±1.00				20.29±0.83	
Change in color	0.00±0.00	10.35±1.92				8.78±1.18	
Mango pulp							Shende and Datta (2020)
L	73.13±0.13	69.74±0.09				55.56±0.11	
a	7.11±0.03	10.86±0.02				14.72±0.02	
b	76.15±0.22	58.64±0.14				41.86±0.27	

					Reference
Change in color	18.22				39.27
Apple slices					Baeghbali et al. (2019)
L	66.03±1.65	55.74±1.65	58.73±1.65		53.35±1.65
a	−15.98±0.58	−12.38±0.58	−14.21±0.58		−12.37±0.58
b	17.52±0.73	12.01±0.73	11.32±0.73		24.92±0.73
Change in color		12.21	9.74		15.12
Dragon fruit peel					Gautam and Abdul (2020)
L	47.34	27.43	32.48		27.97
a	37.39	19.02	22.48		12.56
b	−0.90	6.94	4.86		11.26
Change in color	28.21	21.82			33.76

(Continued)

Table 3.1 (Continued)

Product	Fresh	RWD	DD	FD	SD	OD/TD	References
Guava pulp							Leiton-Ramírez et al. (2020)
L		8.75		21.57		5.39	
a		2.74		−1.67		3.54	
b		11.55		−1.49		12.89	
Change in color		8.46		24.40		18.78	
Banana puree							Rajoria et al. (2021)
L	74.35±2.04	71.66±1.11					
a	3.46±0.59	2.81±0.20					
b	27.46±1.48	25.46±0.40					
Change in color		3.44±1.07					

RWD refers to refractance window drying, DD refers to drum drying, FD refers to freeze drying, SD refers to spray drying, OD refers to oven drying, TD refers to tray drying.

Table 3.2 Comparative Description of the Optimized Drying Conditions of Different Products

Drying conditions	Drying time	Final moisture content (% w.b.) or water activity (Aw)	Energy consumption for drying of 1 kg sample (1 kWh)	Overall energy efficiency (%)	References
Strawberry puree, strawberry puree + maltodextrin, carrot puree					
RWD 95°C of hot water temperature, 0.45 to 0.58 m/min of belt speed, 1 mm of puree thickness	3 to 5 min	6.1 (carrot puree), 5.7 (strawberry puree with 70% maltodextrin as carrier), and 9.9 (strawberry puree without carrier)	—	—	Abonyi et al. (2002)
DD 138°C of pressurized steam	3 min	5.0	—	—	
FD Samples were quick-frozen at –35°C, absolute pressure of 3.3 kPa, 20°C of heating plate temperature, –64°C of condenser temperature	24 h	8.2 (carrot puree), 3.9 (strawberry puree with 70% maltodextrin as carrier), and 12.1 (strawberry puree without carrier)	—	—	
SD Inlet air temperature was 190±5°C and the outlet air temperature 95±5°C.	—	2.3	—	—	

(Continued)

Table 3.2 (Continued)

	Drying conditions	Drying time	Final moisture content (% w.b.) or water activity (Aw)	Energy consumption for drying of 1 kg sample (1 kWh)	Overall energy efficiency (%)	References
Tomato juice, tomato ketchup, carrot puree						Baeghbali et al. (2010)
RWD	95°C of hot water temperature	5–7 minutes	5.7	4.5	–	
FD	–	20–24 h	7.2	3.5	–	
Mango powder						Caparino et al. (2012)
RWD	95–97°C of hot water temperature, 0.5–0.7 mm of puree thickness	180±0.15 seconds	4.76	–	–	
DD	152±2°C, steam pressure 379.2±7 kPa	54±0.2 seconds		–	–	
FD	Samples were quick-frozen at −25°C, vacuum pressure of 20Pa, 20°C of heating plate temperature, −60°C of condenser temperature	111,600±5091 seconds		–	–	

					Reference
SD	Inlet air temperature was 190±2°C and the outlet air temperature 90 ±2°C	1–3 seconds	—	—	—
Pomegranate concentrate					
RWD	91°C of hot water temperature, 20°C of cooling water temperature, 3.9 mm/s of Mylar belt speed, 0.5 mm thickness of pomegranate concentrate	8.5 minutes	5.38±0.57	4.31 ± 0.82	31.56	Baeghbali et al. (2016)
FD	Samples were quick-frozen at −80°C, absolute pressure of 3.0 kPa, 20°C of heating plate temperature, −40°C of condenser temperature	24 h	8.55±0.57	130.65 ± 0.82	1.12
SD	The inlet and outlet air temperatures were 140±1°C and 75±1°C, respectively	—	2.92±0.57	11.01 ± 0.82	12.92

(Continued)

Table 3.2 (Continued)

	Drying conditions	Drying time	Final moisture content (% w.b.) or water activity (Aw)	Energy consumption for drying of 1 kg sample (1 kWh)	Overall energy efficiency (%)	References
Apple slices						
RWD	95±1°C of water temperature, 2 mm thick of plastic film (Mylar, polyethylene tetraphthalate), 4 mm thick and 40-mm-diameter apple slices	50 minutes	Aw = 0.4	–		Hernández et al. (2019)
TD	55±1°C of drying temperature, 2.23 m/s of average air rate	>270 minutes	Aw = 0.4	–		
Mango pulp						
RWD	95±2°C of hot water temperature, 0.15 mm thick of Mylar film, 2.50 mm thickness of peel pulp	34.67±0.57 minutes	17.43±0.05			Shende and Datta (2020)

Method	Conditions			Reference
TD	95°C of drying temperature, 2.50 mm thickness of peel pulp	258.33±10.41 minutes	18.58±0.06	Gautam and Abdul (2020)
OD	95°C of drying temperature, 2.50 mm thickness of peel pulp	441.67 ± 7.64	18.85±0.06	

Dragon peel

Method	Conditions			
RWD	95°C of hot water temperature, 0.15 mm thick of Mylar film, 3.5 mm thickness of peel pulp	42 minutes	5–6	
FD	Samples were quick-frozen at −40°C, freeze dryer at −50°C under 0.110 mBar pressure	72 h	5–6	
TD	70°C of drying temperature, 1 m/s of average air rate	480 minutes	5–6	

(Continued)

Table 3.2 (Continued)

Drying conditions	Drying time	Final moisture content (% w.b.) or water activity (Aw)	Energy consumption for drying of 1 kg sample (1 kWh)	Overall energy efficiency (%)	References
Guava pulp					
RWD 90°C of hot water temperature, 2 mm thickness of peel pulp	76 minutes	3.85			Leiton-Ramirez et al. (2020)
FD Samples were quick-frozen at −35°C, freeze dryer at −35°C under 8 Pa pressure	107 minutes	3.85			
TD 60±0.5°C of drying temperature, 0.6±0.02m/s of air velocity	240 minutes	3.85			

RWD refers to refractance window drying; DD refers to drum drying; FD refers to freeze drying; SD refers to spray drying; OD refers to oven drying; TD refers to tray drying.

3.6 Industrial Applications in the Food Sector

The most preferred and popular drying methods for food processing employed in an industrial scale are cabinet/tray dryers, drum dryers, vacuum dryers, spray dryers, and fluidized bed dryers. (Parikh 2014). The drying process can take many forms and utilizes different types of dryers, with each developed to suit a given operation or product. Jayaraman and Das Gupta (1992) observed that the increasing rejection of food based on quality and the demand for a sustainable and comprehensive range of products has generated renewed interest in drying operations. Moreover, consumer choice has been raised towards the nutritional, functional, and sensory aspects of food products with an objective of retaining the maximum bioactive compounds. Hence, processors and industrialists have become more precise in selecting the types of dryers that are tailored for specific food products.

The application of RW drying in the food industry is anticipated to represent a new process that can produce cost-effective high-quality dried foods (Mahanti et al. 2021). Some of the studies on RW dryers give better insights into the technology and its applications in different segments of food industries.

3.6.1 Fruit and Vegetable Processing Industry

RW drying has been successfully used for drying of a variety of horticultural produce to obtain value-added products like puree, slices, juices, powder, flakes, and sheets with retention of quality and heat-sensitive nutritional components (Nindo and Tang 2007; Karadbhajne et al. 2019). MCD Technologies initially created the term refractance window, which features a short drying time, typically 3–5 min and 50–70% lower capital investment and is 50% more energy efficient than freeze dryers. The temperature reached within the product is lower than 70°C (Magoon 1986). RW dryers are suitable to produce juice, fruit leather, and powder of soup and stews at the industrial scale. Some reports have proved the efficiency of RW in retaining food product qualities in contrast with conventional drying methods. Specific studies were conducted on strawberry, mango, and asparagus in retaining color and total antioxidants and preserving β-carotene in carrots over drum dryers (Caparino 2012; Nindo and Tang 2007). RW-concentrated pomegranate showed higher anthocyanin color and anthocyanin content compared to spray drying (Baeghbali et al. 2016). Cadwallader et al. (2010) reported considerable retention of orange oil using RW drying (75.70%) compared to spray drying (56.90%), with the minimum presence of detrimental components such as limonene oxide. RW-dried tomato powder maintained the antioxidant profile such as total flavonoids, ascorbic acid, and lycopene content over conventional dryers (Abul-Fadl and Ghanem 2011), and other studies showed that quality was maintained in paprika (Topuz et al. 2009), acai juice, (Pavan et al. 2012), and berry juice (Nindo et al. 2004) using RW dryers.

3.6.2 Dairy Industry

Aragón-Rojas et al. (2019) successfully encapsulated the microorganism probiotic *Lactobacillus fermentum* K73 using refractance window drying. The dryer significantly decreased the peroxide value of concentrated milk and increased the shelf life by completely eliminating microorganisms (Al-Hilphy et al. 2021). It can also be an alternative technology for obtaining yogurt powder with enhanced physical properties, thus maintaining the viability of yogurt bacteria (Tontul et al. 2018a).

3.6.3 Meat and Egg Processing Industry

Rostami (2017) prepared meat powder by RW dryer which had good physical properties, high microbial tolerance, negligible Enterobacteriaceae population, and the requisite organoleptic attributes. It was also suggested that even under high RH conditions, RW dried meat powder had low hygroscopicity; therefore, it is desirable to use the technology in lines of food production. Conductive hydro-drying, a kind of RW drying, was found to be an energy efficient, cost-effective, and superior method for drying egg whites with excellent solubility, protein content, powder characteristics, and foaming properties (Preethi et al. 2020).

3.6.4 Nutraceutical Industries

Refractance window drying was reported to retain the quality of functional and nutraceutical foods while preserving the color as well as heat-sensitive vitamins (C & E) (Ayala-Aponte et al. 2021), protein isolates from chickpea (Tontul et al. 2018c), and aloe vera (Link et al. 2017). Encapsulation of orange oil, essential fatty acids, probiotics, flavonoids, organic acids, phenolic compounds, and natural food colorants (Cadwallader et al. 2010; Tontul and Topuz 2017, 2019) from different food extracts, herb extracts was performed using RW drying. This suggested that RW drying has the potential to be used for nutritional and therapeutic applications (Jones 2006).

3.7 Advantages and Limitations

The RW technique, unlike other innovative drying methods, does not require high operational expenses, and the dried products have improved shelf life (**Mt Capra 2017**). The additional advantages of this novel drying technology are low-temperature drying at a lower atmosphere, making it cost-effective, lowering drying time, and maintaining the product's low temperature with minimum chances of cross-contamination. The RW dryer setup is simple in design, consumes less energy, has less maintenance, and is relatively inexpensive, whereas it is higher in thermal efficiency due to the recirculation of hot water back into the reservoir. The product quality is reported to be exceptional compared to the traditional drying techniques like drum drying, freeze drying, vacuum drying, cabinet/tray drying, and spray drying (Bernaert et al. 2019; Smidt et al. 2017; Nindo and Tang 2007).

The limitations of RW drying are reported to be that a mismatch of refractive indices caused due to temperature variation between water, plastic film, and food during processing may stop the drying of food materials (Nindo and Tang 2007; Yoha et al. 2019). Ortiz-Jerez et al. (2015) established that the creation of an air gap between the Mylar film and the product was the outcome of product shriveling and convective cooling while RW drying. Other limitations of RW drying are the large area occupied by the film, product thickness, and the lack of scale-up. The prime limitation in expanding or upscaling is the requirement of a widespread area for drying and exchange of thermal energy. Also, the lower film thickness decreases the handling and processing capacity (Mahanti et al. 2021). RW drying has modest throughput, and its primary application of drying liquids makes it difficult to handle powdery material with high sugar content because of its inherent stickiness.

3.8 Future Perspectives and Conclusion

RW drying has been predicted to have a bright future due to its potential, utility, and possibility of harnessing superior quality and drying of food supplements while maintaining safety standards. Many studies have concluded that the RW is a gentle drying process with excellent nutrient achievement. Compared to other novel dryers, maximum color, texture, and flavor in the final product are retained. Due to its simple and mechanical ease of work, RW drying can be deployed with meager resources in design and development, whereas novel dryers such as freeze dryers and spray dryers essentially need high levels of engineering and technical intervention for their fabrication. RW drying has been demonstrated to be the most cost-effective approach for excellent retention of beneficial compounds when compared with the working principles of recent drying techniques involving higher drying cost. The final product quality is equivalent to spray- or freeze-dried product, yet the equipment cost is substantially less than the setup of spray drying or lyophilization. Better energy conservation can be achieved in the drying process by using solar energy or any other sustainable means to boil the water in the dryer. A temperature up to 90°C used in RW drying may be considered low-quality energy output by a power plant or other industrial facility, and this waste heat could be reutilized in the RW plant. Waste heat obtained from a power plant or any other industrial process could be reutilized in an RW dryer, reducing the energy strain in the industries. Besides solving the post-harvest losses problem, value-added products with extended shelf life can boost the nutritional quality of food and the nutritional status of people. Further investigations need to continue on changes in molecular structure or weight, antioxidant retention, shelf stability of fruit and food polysaccharides processed by this technology, and energy conservation.

References

Abonyi BI, Feng H, Tang J, Edwards CG, Chew BP, Mattinson DS. (2002). Quality retention in strawberry and carrot purees dried with refractance window TM system. *Journal of Food Science*, 67(3): 1051–1056.

Abul-Fadl MM, Ghanem TH. (2011). Effect of refractance-window (RW) drying method on quality criteria of produced tomato powder as compared to the convection drying method. *World Applied Sciences Journal*, 15(7): 953–965.

Al-Hilphy AR, Ali HI, Al-IEssa SA, Lorenzo JM, Barba FJ, Gavahian M. (2021). Refractance window (RW) concentration of milk-part II: Computer vision approach for optimizing microbial and sensory qualities. *Journal of Food Processing and Preservation*: e15702.

Aragón-RojasS,Quintanilla-CarvajalMX,Hernández-SánchezH,Hernández-Álvarez AJ, Moreno FL. (2019). Encapsulation of *Lactobacillus fermentum* K73 by Refractance Window drying. *Scientific Reports*, 9: 5625.

Ayala-Aponte AA, Cárdenas-Nieto JD, Tirado DF. (2021). Aloe vera gel drying by refractance window®: Drying kinetics and high-quality retention. *Foods*, 10: 1445.

Azizi D, Jafari SM, Mirzaei H, Dehnad D. (2016). The influence of refractance window drying on qualitative properties of kiwifruit slices. *International Journal of Food Engineering*, 13(2).

Baeghbali V, Niakosari M, Farahnaky A. (2016). Refractance window drying of pomegranate juice: Quality retention and energy efficiency. *LWT—Food Science and Technology*, 66: 34–40.

Baeghbali V, Niakosari M, Kiani M. (2010). Design, manufacture and investigating functionality of a new batch refractance window system.

Proceedings of 5th International Conference on Innovations in Food and Bioprocess Technology: 1–7.

Baeghbali V, Niakousari M, Ngadi MO, Hadi Eskandari M. (2019). Combined ultrasound and infrared assisted conductive hydro-drying of apple slices. *Drying Technology*, 37: 1793–1805.

Bernaert N, Droogenbroeck BV, Pamel EV, Ruyck HD. (2019). Innovative refractance window drying technology to keep nutrient value during processing. *Trends in Food Science & Technology*, 84: 22–24.

Cadwallader KR, Moore JJ, Zhang Z, Schmidt SJ. (2010). Comparison of spray drying and refractance window™ drying technologies for the encapsulation of orange oil, pp. 246–254. Paper presented at 12th International Flavor Conference—4th George Charalambous Memorial Symposium 2009, Skiathos, Greece.

Calín-Sánchez Á, Lipan L, Cano-Lamadrid M, Kharaghani A, Masztalerz K, Carbonell-Barrachina ÁA, Figiel A (2020). Comparison of traditional and novel drying techniques and its effect on quality of fruits, vegetables and aromatic herbs. *Foods*, 9(9): 1261.

Caparino OA, Tang J, Nindo CI, Sablani SS, Powers JR, Fellman JK. (2012). Effect of drying methods on the physical properties and microstructures of mango (Philippine 'Carabao' var.) powder. *Journal of Food Engineering*, 111(1): 135–148.

Celli GB, Khattab R, Ghanem A, & Brooks MSL. (2016). Refractance window™ drying of haskap berry—preliminary results on anthocyanin retention and physicochemical properties. *Food Chemistry*, 194: 218–221.

da Costa RDS, da Cruz Rodrigue AM, Laurindo JB, da Silva LHM. (2019). Development of dehydrated products from peach palm–tucupi blends with edible film characteristics using refractive window. *Journal of Food Science and Technology*, 56: 560–570.

Durigon A, de Souza PG, Carciofi BAM, Laurindo JB. (2016). Cast-tape drying of tomato juice for the production of powdered tomato. *Food and Bioproducts Processing*, 100: 145–155.

Gautam A, Abdul S. (2020). Comparative study of different drying techniques of dragon fruit peel. *Journal of Pharmacognosy and Phytochemistry*, 9(4): 280–283.

Hernández Y, Ramírez C, Moreno J, Núñez H, Vega O, Almonacid S, Pinto M, Fuentes L, Simpson R. (2019). Effect of refractance window on dehydration of osmotically pretreated apple slices: Color and texture evaluation. *Journal of Food Process Engineering*: 1–11.

Jafari SM, Azizi D, Mirzaei H, Dehnad D. (2016). Comparing quality characteristics of oven-dried and refractance window-dried kiwifruits. *Journal of Food Processing and Preservation*, 40(3): 362–372.

Jalgaonkar K, Mahawar MK, Vishwakarma RK, Shivhare US, Nambi VE. (2020). Optimization of process condition for preparation of sapota bar using refractance window drying method. *Drying Technology*, 38(3): 269–278.

Jayaraman KS, Das Gupta DK. (1992). Dehydration of fruits and veg-etables—recent developments in principles and techniques. *Drying Technology*, 10(1): 1–50.

Jones K. (2006). Dehydration of Food Combinations. US Patent US 2006/0112584 A1.

Karadbhajne SV, Thakare VM, Kardile NB, Thakre SM. (2019). Refractance window drying: An innovative drying technique for heat sensitive product. *International Journal of Recent Technology and Engineering*, 8(4): 1765–1771.

Leiton-Ramirez TM, Ayala-Aponte A, Ochoa-Martinez CI. (2020). Physicochemical properties of guava snacks as affected by drying technology. *Processes*, 8(106): 1–12.

Link JV, Tribuzi G, Laurindo JB. (2017). Improving quality of dried fruits: A comparison between conductive multi-flash and traditional drying methods. *LWT—Food Science and Technology*, 84: 717–725.

Magoon R. (1986). Method and apparatus for drying fruit pulp and the like. US Patent 4, 631, 837.

Mahanti NK, Chakraborty SK, Sudhakar A, Verma DK, Shankar S, Thakur M, Singh S, Tripathy S, Gupta AK, Srivastav PP. (2021). Refractance Window™-drying vs. other drying methods and effect of different process parameters on quality of foods: A comprehensive review of trends and technological developments. *Future Foods*, 3: 100024.

Minjares-Fuentes R, Rodríguez-Gonz´alez VM, Gonz´alez-Laredo RF, Eim V, Gonz´alez-Centeno MR, Femenia A. (2017). Effect of different dry-ing procedures on the bioactive polysaccharide acemannan from Aloe vera (*Aloe barbadensis* Miller). *Carbohydrate Polymers*, 168: 327–336.

Mt Capra (2017). www.mtcapra.com/refractance-window-drying technol-ogy. Accessed on 22 August 2021.

Nayak B, Berrios JDJ, Powers JR, Tang J, Ji Y. (2011). Coloured potatoes (*Solanum tuberosum* L.) dried for antioxidant-rich value-added foods. *Journal of Food Processing and Preservation*, 35: 571–580.

Nemzer B, Vargas L, Xia X, Sintara M, Feng H. (2018). Phytochemical and physical properties of blueberries, tart cherries, strawberries, and cranberries as affected by different drying methods. *Food Chemistry*: 242–250.

Nindo CI, Tang J. (2007). Refractance window dehydration technology: A novel contact drying method. *Drying Technology*, 25(1): 37–48.

Nindo CI, Tang J, Powers JR, Bolland K. (2004). Energy consumption during refractance window evaporation of selected berry juices. *International Journal of Energy Research*, 28: 1089–1100.

Ochoa-Martinez CI, Quintero PT, Ayala AA, Ortiz MJ. (2012). Drying char-acteristics of mango slices using the Refractance Window™ technique. *Journal of Food Engineering*, 109: 69–75.

Ortiz-Jerez MJ, Gulati T, Datta AK, Ochoa-Martínez CI. (2015). Quantitative understanding of refractance window™ drying. *Food and Bioproducts Processing*, 95: 237–253.

Ortiz-Jerez MJ, Ochoa-Martinez CI. (2015). Heat transfer mechanisms in conductive hydro-drying of Pumpkin (*Cucurbita maxima*) pieces. *Drying Technology*, 33(8): 965–972.

Papoutsis K, Vuong QV, Golding JB, Hasperue JH, Pristijono P, Bowyer MC. (2018). Pretreatment of citrus by-products affects polyphenol recovery: A review. *Food Reviews International*, 34(8): 770–795.

Parikh DM. (2014). Solids drying: Basics and applications. *Chemical Engineering-New York-McGraw Hill Incorporated then Chemical Week Publishing LLC*, 121(4): 42–45.

Pavan MA, Schmidt SJ, Feng H. (2012). Water sorption behaviour and thermal analysis of freeze-dried, Refractance Window-dried and hot-air dried açaí (*Euterpe oleracea Martius*) juice. *LWT Food Science and Technology*, 48: 75–81.

Preethi RD, Shweta D, Moses JA, Anandharamakrishnan C. (2020). Conductive hydro drying as an alternative method for egg white powder production. *Drying Technology*. Doi: 10.1080/07373937.2020.1788073.

Qiu J, Acharya P, Jacobs DM, Boom RM, Schutyser MAI. (2019). A systematic analysis on tomato powder quality prepared by four conductive drying technologies. *Innovative Food Science & Emerging Technologies*, 54: 103–112.

Raghavi LM, Moses JA, Anandharamakrishnan C. (2018). Refractance window drying of foods: A review. *Journal of Food Engineering*, 222: 267–275.

Rajoria D, Bhavya ML, Hebbar U. (2021). Impact of process parameters on drying behaviour, mass transfer and quality profile of refractance window dried banana puree. *LWT—Food Science and Technology*, 145: 111330.

Rajoriya D, Shewale SR, Hebbar HU. (2019). Refractance window drying of apple slices: Mass transfer phenomena and quality parameters. *Food and Bioprocess Technology*, 12(10): 1646–1658.

Ratti C, Mujumdar AS (1995). Infrared drying. In Mujumdar AS (Ed.), *Handbook of Industrial Drying*, Marcel Dekker: New York, pp. 567–588.

Rostami H, Dehnad D, Jafari SM, Tavakoli HR. (2017). Evaluation of physical, rheological, microbial and organoleptic properties of meat powder produced by refractance-window drying. *Drying Technology*. Doi: 10.1080/07373937.2017.1377224.

Shende D, Datta AK. (2020). Optimization study for refractance window drying process of Langra variety mango. *Journal of Food Science and Technology*, 57(2): 683–692.

Smidt RP, Wemmers AK, Spoelstra S. (2017). Thin film drying processes energy and economic aspects compared to spray drying. (Project report registered at Energy Research Centre of the Netherlands (ECN) under project number 5.3615), 2017.

Tontul I, Ergin F, Eroğlu E, Küçükçetin A, Topuz A. (2018a). Physical and microbiological properties of yoghurt powder produced by refractance window drying. *International Dairy Journal*, 85: 169–176.

Tontul I, Eroglu E, Topuz A. (2018b). Convective and refractance window drying of cornelian cherry pulp: Effect on physicochemical properties. *Journal of Food Process Engineering*, 41(3): e12917.

Tontul I, Kasimoglu Z, Asik S, Atbakan T, Topuz A. (2018c). Functional properties of chickpea protein isolates dried by refractance window drying. *International Journal of Biological Macromolecules*, 109: 1253–1259.

Tontul I, Topuz A. (2017). Effects of different drying methods on the physicochemical properties of pomegranate leather (pestil). *LWT—Food Science and Technology*, 80: 294–303. Doi: 10.1016/j.lwt.2017.02.035.

Tontul I, Topuz A. (2019). Storage stability of bioactive compounds of pomegranate leather (pestil) produced by refractance window drying. *Journal of Food Process Engineering*, 42(2): e12973.

Topuz A, Feng H, Kushad M. (2009). The effect of drying method and storage on color characteristics of paprika. *LWT—Food Science and Technology*, 42: 1667–1673.

Verma DK, Thakur M, Srivastav PP, Karizaki VM, Suleria HAR (2020). Effects of drying technology on physiochemical and nutritional quality of fruits and vegetables. In Srivastav PP, Verma DK, Patel AR, Al-Hilphy AS (Eds.), *Emerging Thermal and Nonthermal Technologies in Food Processing*. Apple Academic Press, Florida, USA, pp. 69–116.

Yoha KS, Priyadarshini SR, Moses JA, Anandharamakrishnan C. (2019). Refractance window drying and its applications in food processing. 1st ed. In Deka SC, Seth D, Hulle S (Eds.), *Technologies for Value Addition in Food Products and Processes*, Apple Academic Press, Taylor & Francis: New York, pp. 61–72.

Zia S, Khan MR, Shabbi, MA, Maan AA, Khan MKI, Nadeem M. (2020). An inclusive overview of advanced thermal and nonthermal extraction techniques for bioactive compounds in food and food-related matrices. *Food Reviews International*: 1–31.

Chapter 4

Ozone Technology in Food Disinfection

Dolly Bhati, Arashdeep Singh, and Gurkirat Kaur

4.1 Introduction

One of the most important fundamentals for the endorsement of safety and quality management in food processing industries is the competent inactivation of spoilage and disease-causing microorganisms from various food commodities. Raw supplies and semi, minimally, and completely processed goods consisting of cereal, pulses, oilseeds, meat, fish, poultry, fruits, and vegetables are of prime concern. Currently, given revolutionary developments in the domains of food processing and preservation, conventional approaches such as thermal processing are becoming obsolete. On the one hand, owing to a number of significant shortcomings such as negative impacts on the sensorial and nutritional characteristics of products, the conventional techniques of thermal food processing are not prominent these days. On the other hand, contemporary approaches of non-thermal food processing not only prevent nutritional losses but also ensure the conservation of fundamental flavor, taste, appearance, color, and texture. The industrial application of ozone is not new; it has been employed for many years in a number of domains such as water treatments and as a disinfectant for safeguarding and extending the shelf life of products during the course of storage. In recent times, the utilization and importance of ozone has become diversified in food processing and preservation.

DOI: 10.1201/9781003217138-4

83

The special qualities of ozone processing, namely its quick and efficient inactivation of microorganisms and lack of production of dangerous residues, suggest that ozone is safe to use in food processing. If employed precisely, ozone can not only eradicate the damaging effects of thermal processing such as deterioration of organoleptic characteristics; it can also ensure an effluent or treatment medium free of hazardous residues. Owing to the remarkable oxidation potential of ozone, it is a prominent disinfectant that guarantees its extensive utility in ensuring the quality and safety of food. Disinfecting agents, for instance chlorine, are categorically ineffective and uneconomical when it comes to the inactivation of heat-resistant spore-forming bacteria under high pH conditions. In addition, trihalomethanes have a predisposed tendency to react and form complexes with chlorine molecules, thereby resulting in the formation of compounds that are of vital concern for both environmental and human dietary safety.

On account of these considerations, the present-day modernizing food processing industry is in pursuit of disinfecting and preservative agents that are:

- Cost effective and useful in exterminating customary spoilage and disease-causing microorganisms, along with hazardous contaminants.
- Able to ensure negligible detrimental influences on the product's nutritional (limiting amino acids, antioxidants, phytochemicals, etc.) and organoleptic (texture, flavor, aroma, taste) excellence.
- Versatile enough to facilitate broad-spectrum compliance with different food processes.
- Able to have a minimum detrimental influence on the environment.

It has been demonstrated that ozone has bactericidal properties against a wide range of microorganisms, including yeast, molds, bacteria with varying morphologies (coccus, bacillus, vibrio, and spirilla), cell wall composition (Gram-positive and Gram-negative), and oxygen requirements (aerobic, facultative, or anaerobic). Similarly, frequent industrial uses of ozone in diverse processes such as hygienic maintenance of food contact surfaces, plant equipment disinfection, decontamination of waste water, and reduction of chemical and biological oxygen demands of effluent generated from food processing plants has also been established over time. Ozone is not only able to be employed as a prominent food disinfectant in developed nations; its use is gaining extensive popularity in developing nations as well. This chapter attempts to describe the physical and chemical characteristics of ozone, techno-functional aspects of microbial inactivation, target locations for disinfection activity, aids in creating complimentary associations with ozone molecules to enhance its disinfection potential, contemporary functions of ozone in the area of food processing industries, and technical glitches often encountered in the implementation of ozone processing in food industries, along with credible and interesting upcoming applications.

4.2 Ozone: Physicochemical Outline

In 1840, a scientist named Christian Friedrich Schönbein perceived a strange odor in his laboratory while performing his experiments. This unique odor gave him an idea regarding the existence of a novel compound. Owing to the intensity of the odor, he coined the name "ozone" for the novel gas, adapted from the Greek term "ozein", signifying "smell" (Galdeano *et al.*, 2018). When a free radical of oxygen (O.) bonds with molecular oxygen (O_2), it results in the generation of a tri-atomic oxygen entity known as ozone (O_3). The middle oxygen atom is linked with the two other oxygen atoms with a bond length of 1.278 Å and is equidistant from both atoms. The entire atomic assembly is positioned at an obtuse angle on the order of 116° 492. Several other critical thermodynamic characteristics of ozone molecule are a 111.9° ± 0.3°C boiling point, 192.5° ± 0.4°C melting point, 12.1°C critical temperature, and 54.6 atm critical pressure. The density of gaseous ozone, 2.14 gL^{-1} at 0°C and 101.3 kPa, is superior to that of air (1.28 gL^{-1}) under equivalent conditions. The natural synthesis of ozone is credited to the shorter wavelength with ultraviolet solar radiation, usually with a wavelength less than 240 nm. In the stratosphere as high as 20 km, when this ultraviolet radiation strikes oxygen molecules (O_2), it results in splitting it off into two oxygen atoms (O.), which subsequently associate with oxygen molecules and result in the synthesis of ozone (O_3). The synthesized ozone assimilates itself well within the stratosphere and troposphere. Ozone is an allotropic amendment of native oxygen molecules. In contrast to the molecular weight (32.00) of a diatomic oxygen molecule, the molecular weight of an ozone molecule is 48.00. The molecular structure of ozone is postulated to be the resonance amalgam of the four canonical forms, demonstrating delocalized bonding. Ozone gas and liquid possess a characteristic soft blue hue and a distinctive peculiar smell comparable to the aroma of air immediately after a thunderstorm. The only by-product of ozone decomposition is oxygen, which is the prime reason the effluent generated during the course of ozone processing is completely devoid of disinfectant residue. The color characteristics of ozone vary depending on the temperature conditions and purity of substrate employed for the synthesis; for example, when ozone is generated from dried air, it has a light blue color. On the contrary, when ozone is synthesized from high-purity oxygen, it is usually devoid of any hue. At ambient to low storage temperatures, ozone has moderate solubility in aqueous media. The basic trait explaining the superior reactivity potential of ozone is its distinguishable oxidizing potential of 2.07 V, which establishes it as a prominent oxidant comparable to fluorine and persulphate.

4.2.1 Nature of Solubility

There is an wide scale of physical characteristics that have a substantial influence over the solubility characterization of ozone. By virtue of these physical factors, ozone is reasonably soluble in aqueous media and results in the formation of a true solution. Henry's law elucidates the solubility behavior of ozone in water; the law states that "The application of pressure to a vapor phase when it is in equilibrium with a liquid phase is inversely related to the temperature of the system." In accordance with Henry's law, the solubility of ozone in an aqueous medium is immediately relative to the partial pressure exerted by ozone gas above the surface of the aqueous medium. Out of all the conditional factors, temperature is the most crucial one that has a noteworthy impact on the solubility of ozone in water. Consistent with this crucial determining feature and Henry-Dalton constants, it has been postulated that ozone shows appreciable solubility under low-temperature conditions. As stated by Horvath *et al.* (1985), ozone demonstrates an aqueous solubility quotient ranging from 0.31 to 1.13, which is chiefly dependent on the temperature of the aqueous system. Ozone's solubilization behavior diverges in accordance with the origin of water utilized for its solubilization. Kim (1998) carried out a characterization of ozone's solubility in different types of aqueous mediums. For instance, he showed that when ozone was bubbled at a flow rate of 29.4 mL min^{-1} for 18 min in distilled water, it demonstrated a solubility of 16% at approximately 22°C in contrast to 38% and 98% solubility in tap and deionized water. The dissimilarities in the solubility behavior of ozone are credited to the disparities in the analytical method implemented, flow rate employed for sparging of ozone, and design of the reactor. When compared with the solubility of oxygen gas in water, the overall ozone solubility was found to be approximately 13 times greater at 0°C to 30°C, and it was found to be increasingly soluble in colder aqueous systems (Galdeano *et al.*, 2018).

The second most crucial characteristic impacting the solubility of molecular ozone in any medium is pH. High pH conditions (>8) hinder the solubilization of ozone in aqueous media (Jung *et al.*, 2017). Galdeano *et al.* (2018) scrutinized the patterns for concentration and time of ozone saturation in various aqueous systems as impacted by variable temperature and pH conditions. They observed that temperature and pH are the two most vital factors with a critical impact on the concentration of ozone in the aqueous system. Researchers proposed that as these two variables control the varying rates of ozone gas breakdown in diverse media, they have a significant impact on ozone solubility. High temperature and neutral to alkaline pH conditions of the aqueous system result in hydroxyl ions, encouraging prompt breakdown of the gas. Ferriera *et al.* (2017) found a superior survival rate of aerobic mesophiles with an increase in pH and temperature of the treatment medium while establishing the potency of an aqueous ozone system in restricting the number and growth of micro-organisms on stored strawberries, along with substantiating its impact on the overall quality of the fruit. Liu *et al.* (2016) ascertained that hydroxide ions are used up in the process of ozone breakdown in an aqueous system; therefore, ozonation of a solution could be responsible for the decrease in its pH value.

4.2.2 Nature of Stability

While investigating the potential of gaseous ozone as an antifungal fumigant for grains for storage, Wu *et al.* (2006) postulated that when it comes to the stability of ozone, the gaseous state of the disinfectant has superior stability compared to that of the aqueous phase. Analytically, the characteristic most commonly used to determine the stability of any form, that is, gaseous or solubilized, of ozone is half-life. A broad spectrum of factors, including presence of antioxidants—radical scavengers and metal ions (Hirahara *et al.*, 2019), exposure to UV radiation, presence of organic residues, rate and speed of agitation, concentration of oxygen in the medium, chemical composition (source, pH, minerals) (Ferriera *et al.*, 2017), and temperature of aqueous medium, are principally responsible for governing the half-life of ozone (Dolly *et al.*, 2020). Per the study of Alexopoulos *et al.*, (2013) the breakdown of ozone molecules in water does not always progress per first-order rate dynamics.

Ferriera *et al.* (2017) showed that a phosphate buffer with pH ≥ 8 catalyzes rapid disintegration of ozone. They observed the thermodynamics of ozone disintegration spectrophotometrically in different purity grades of water such as tap water, distilled water, deionized water, and High-performance liquid chromatography HPLC-grade water that were sparged with a stream of analytical-grade ozone at a constant flow rate. They found that the stability of ozone was in order of HPLC-grade water > deionized water > distilled water > tap water > phosphate buffer. They reported that less than 10% of the ozone was dissociated in the double-distilled water at 20°C even after 85 minutes, whereas the disintegration was much faster in the cases of tap and distilled water, where more than 50% of the ozone was dissociated within duration of 20 minutes.

In 2021, Psaltou *et al.* (2021) assessed the role of transition metal ions as ozonation catalysts and as a potential substitute for the conventional approach of heterogeneous catalytic ozonation. They concluded that the water system is an important consideration when it comes to the process of ozonation, particularly concerning the use of metal ions as catalysts. Co^{+2} and Fe^{+2} demonstrated satisfactory catalytic action in deionized water and dechlorinated potable water systems. The presence of PO_4^{3-} in minimal concentrations functions as a scavenger, thereby lowering the oxidation capacity of the ozonation system. Temperature acts as a decisive dynamic in predicting ozone's stability in different aqueous media. Galdeano *et al.* (2018) postulated that the half-life of ozone in a gaseous state is nearly 12 hours in comparison to that dissolved in clean potable water with a pH in the range of 7 to 8, which is nearly equivalent to 20 to 30 minutes at ambient temperature. Furthermore, the amount of ozone-consuming materials present in an aqueous system is a vital dynamic that has a noteworthy influence on its half-life.

4.2.3 Nature of Reactivity

4.2.3.1 Reactions of Molecular Ozone

As already established, the significantly high reactivity and poor stability of ozone gas cause it to rapidly dissociate in the environment. The chief reason for the substantially high reactivity of ozone in any aqueous matrix is its highly unsteady molecular electronic configuration. We are aware that an ozone molecule is made up of three oxygen atoms, each of which has two unpaired electrons in its valency shell (Gardoni *et al.*, 2012). Ozone is a molecule that demonstrates a hybrid of four probable resonance structures. The electrophilic nature (tendency to form bonds by accepting an electron pair) of the ozone molecule is characterized by those resonance structures, which show a lack of an electron pair in one of the oxygen atoms at the end of the molecule. On the contrary, additional anionic charge manifested by one of the other two atoms ends up conveying a nucleophilic character to the molecule (Beltrán, 2003). Ozone participates in three different classes of interactions in an organic solvent matrix (Nakano *et al.*, 2003):

- **Category 1** Molecular ozone has a propensity towards dipolar cyclo-addition interactions with unsaturated linkages between two carbon atoms.
- **Category 2** Predisposition towards electrophilic interactions with cyclic compounds, amines, and sulfides with resilient electron density.
- **Category 3** Tendency towards nucleophilic interactions with carbons constituting electron-withdrawing groups.

Owing to these predispositions towards these categories of reactions, the reactions of molecular ozone are not only discerning, they are also simultaneously controlled to unsaturated aromatic and aliphatic compounds along with specific functional groups.

4.2.3.2 Reactions with Decomposition Products—Free Radical Species

There are two modes of interactions possible between ozone dissolved in the aqueous matrix and the respective decomposition products, direct and indirect interactions:

1. *Direct interactions* take into account ozone molecules (O_3) and other molecular entities or compounds such as organic/inorganic molecules and radicals.
2. *Indirect interactions* take into account the hydroxyl radicals ($OH^\cdot$), which are generated by virtue of the molecular decay of ozone and other molecules (Khadhraoui *et al.*, 2009).

Direct interactions of ozone molecules with organic and inorganic molecules present in the aqueous matrix are very specific in nature (Bezbarua and Reckhow, 2004; Fábián 2006; Ignatiev *et al.* 2008). During most of the time, direct oxidation involves nucleophilic carbons, atoms with a negative charge density (e.g., N, P, O), and elements with numerous bonds, such as carbon-carbon and nitrogen-nitrogen (Khadhraoui *et al.*, 2009). Due to the dipolar nature of the ozone molecule, it reacts with the unsaturated carbon bond, separating the bonds according to Criegee mechanism (Beltrán, 2003; Gottschalk *et al.*, 2009). For years, whenever direct interactions between ozone and organic molecules have been investigated, a broad spectrum of variations in the thermodynamics of the reaction's kinetic constants has been witnessed (Von Gunten, 2003).

In the case of indirect interactions, the hydroxyl ($OH^\cdot$) radical and other particular radicals are taken into consideration. Three most common free radicals encountered in ozone-solubilized aqueous systems are hydroperoxyl ($HO_2^\cdot$), superoxide ($O_2^{\cdot-}$), and hydroxyl ($OH^\cdot$). The hydroxyl radical, which has an internal energy of 2.80 V, is one of the most potent, transitory chain reaction-propagating, imperative reactive entities. Owing to this substantially high internal energy, it has a tendency to react vigorously and nonselectively with ozone molecules (Bezbarua and Reckhow, 2004; Fábián, 2006; Ignatiev *et al.*, 2008).

The following discussion examines the molecular degradation of ozone in an aqueous matrix. The mechanism of self-disintegration of molecular ozone in an aqueous system is very complex. In the mid-1980s two different pathways were proposed for the auto-decomposition of ozone. The first pathway for the breakdown of molecular ozone under a neutral to acidic environment was proposed by Staehelin and Hoigné (1982), followed by the second mechanism for the disintegration of ozone under alkaline conditions (in the presence of bicarbonates and carbonates), suggested by Tomiyasu *et al.* (1985).

Fresh reconsiderations of the conventional mechanism for chain reactions were proposed by Bezbarua and Reckhow (2004) and Igantiev *et al.* (2008). Elucidation of the mechanism for ozone disintegration is based on the chain reaction series starting from the steps of initiation, propagation, and termination (Gardoni *et al.*, 2012; Ferriera *et al.*, 2017; Psaltou *et al.*, 2021)

1. *Initiation*: The molecular entities capable of inducing the generation of free radicals are designated as initiators, and the class of reactions is designated as initiation reactions. For instance, hydroperoxide ions (HO_2^-), superoxide radical ($^\cdot O_2^-$), hydroxyl ions (OH^-), and some cations and organic molecules (e.g. humic substances, formic acid, and glyoxylic acid) are recognized as initiators. Naturally ultraviolet radiation with a wavelength of 253.7 nm also prompts the lysis of ozone and results in the formation of free radicals.
2. *Propagation*: Chemical reactions ensuring reinforcement of superoxide radicals from the hydroxyl radicals are designated as propagation reactions. Primary alcohols, aryl groups, and inorganic phosphate species, along with glyoxylic and humic acids, are classified as the most common propagators that encourage ozone degradation.
3. *Inhibition*: Inhibition reactions are defined as the use of hydroxyl radicals without the regeneration of peroxide radicals. For instance, some of the customary inhibitors involve molecular entities such as humic substances, bicarbonate and carbonate ions, alkyl groups, and tertiary alcohols.

The reactivity potential of ozone is chiefly controlled by the oxidation strength of its radical species. On the other hand, naturally occurring antioxidants such as vitamins E and C present in food matrixes can competently hunt the radical species, thereby terminating the string of reactions.

4.2.3.3 Reactions with Inorganic Molecules

Chemical entities such as various minerals, metal cations, hydroxyl ions, and halogenic anions such as chlorine (Cl^-) and fluorine (F^-) catalyze the breakdown of ozone in a very efficient manner, thereby successively increasing the ozone demand. Unlike the previously discussed interactions, that between molecular ozone and inorganic molecules present in the aqueous matrix categorically follows a first-order kinetics. Some of the usual interactions of ozone are with iron and manganese species:

1. Ferrous species (Fe^{2+}) → Ferric species (Fe^{3+}); ozone reacts with the soluble form of iron and results in the formation of an insoluble precipitate, ferric hydroxide, $Fe(OH)_3$, that can be filtered without any difficulty by the process of filtration.
2. Manganese species (Mn^{2-}) → Manganic species (Mn^{4-}); in an analogous manner, ozone reacts with the soluble form of manganese and generates an insoluble precipitate, manganic oxide (MnO_2), that can be filtered with ease by the process of filtration.

These reactions are vital for the ozone-assisted purification of water that is contaminated with metal ions. Owing to the characteristic variations among the redox potentials of ozone O_3 (2.07 V), chloride Cl– (1.49 V), bromide Br – (1.33 V), and iodide I – (0.99 V), ozone can efficiently react with all the respective molecular entities, although the speed of chemical reactions differs significantly according to the oxidation-reduction potential of the species involved. For instance, the speed of ozone-assisted oxidation of chloride ions is very slow, followed by moderate oxidation of bromide ions and prompt oxidation of iodide ions.

4.3 Generation of Ozone

As discussed in the previous sections, ozone is a three-atom–containing allotrope of oxygen, and owing to its substantially high reaction potential, it is very unstable and tends to break down into oxygen naturally. Consequently, for industrial purposes, ozone needs to be synthesized at the point of use. There are numerous methods available for synthesis of ozone at the industrial scale, with a broad spectrum of advantages, disadvantages, and specific industrial applicability.

4.3.1 Dielectric Barrier Discharge (Corona Discharge)

This is one of the most economically significant procedures employed for the production of ozone at an industrial scale. It is also known as the "silent electrical discharge" method. The dielectric barrier discharge method is somewhat analogous to the process of synthetic lightning production. The entire assembly consists of a two-electrode system with a space for the

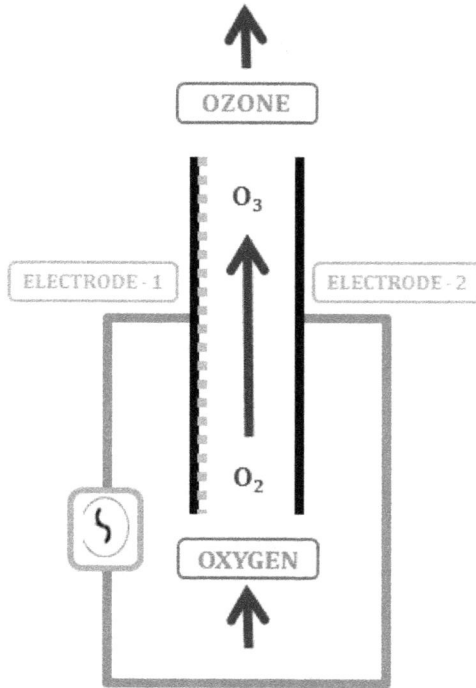

Figure 4.1 Assembly for dielectric barrier discharge (corona discharge).

movement of an air steam (oxygen, moisture-free air, or a mixture thereof) and an insulating dielectric barrier coated on one of the electrodes along with an alternating high-voltage current device, represented in Figure 4.1. Out of the two electrodes, one is designated as a high-energy electrode, while the other is a ground electrode. These electrodes are separated by a narrow space filled by a dielectric medium where the energy-intensive atomic and molecular collisions take place.

The basic working mechanism of the dielectric barrier discharge system takes into account the application of a high-voltage alternating current (approximately equivalent to ~10 kV) through the electrodes that prompts the lysis of oxygen molecules into high energy–manifesting individual oxygen atoms. Then these reactive oxygen species interact with stable oxygen molecules by virtue of energy-intensive multiple collisions, thereby resulting in the formation of ozone. These high-energy atomic and molecular collisions between the oxygen molecules and the reactive species ensure the generation of one molecule of ozone per molecule of oxygen. A dielectric barrier discharge becomes apparent when the injected stream of air becomes moderately ionized and as a consequence has distinctive violet hue, specifically when the feed to the system is moisture-free air. But when high-purity oxygen is employed as the feed to the system, the violet coloration is absent.

The efficiency of ozone formulation by corona discharge is subject to a broad spectrum of elements, such as the dimensions of the discharge space, in-feed oxygen pressure, characteristics of the high-tension and dielectric-coated ground electrode, power input, and humidity. In low-intensity discharge systems, a substantial proportion of the power input is utilized by ions, although in high-intensity discharge system, virtually the entire energy of discharge is conveyed to the electrons responsible for the generation of O_3. If normal air is employed as a feed to the discharge unit, it will essentially be dehydrated and devoid of traces of heavier substances because they are susceptible to ozone-stimulated oxidation. The presence of moisture in the air results in the generation of nitrogen oxides in a generator assembly that will produce nitric acid, which is corrosive in nature, and gradually corrode the generator, necessitating repeated upkeep and down time. The utilization of highly purified oxygen results in ozone yield as high as 16% in comparison to a conventional air feed, which results in 1–3% ozone. This method of ozone generation is the most widely used.

4.3.2 Photochemical or Ultraviolet Radiation–Induced Ozone Production

The basic principle behind the photochemical generation of ozone is comparable to what happens in the stratosphere: singlet oxygen atoms (O) produced by ultraviolet radiation (short wavelength) stimulate the breakdown of oxygen molecules, which subsequently interact with native oxygen moieties (O_2) to produce ozone (O_3). Despite the fact that this mechanism is thought to produce only 2% of ozone. However, practically, the genuine fixed output of ozone is reduced by a factor of approximately 0.5%. For this reason, mercury lamps (operating on low pressure) employed in the photochemical production of ozone do not merely generate the 185-nm radioactive waves that are responsible for the fabrication of ozone but also produce a considerable fraction of UV rays with 254-nm wavelength that are capable of destroying it. While mercury lamps running at medium to high pressure generate excellent UV light with a wavelength of 185 nm that promotes increased ozone generation. One of the most prominent benefits of UV radiation–stimulated photochemical generation of ozone is that normal air can be employed as the input stream.

In spite of all these features, it is not usually possible to produce substantially high intensities of ozone exclusively with a UV-based photochemical production method. Low intensities of ozone produced from

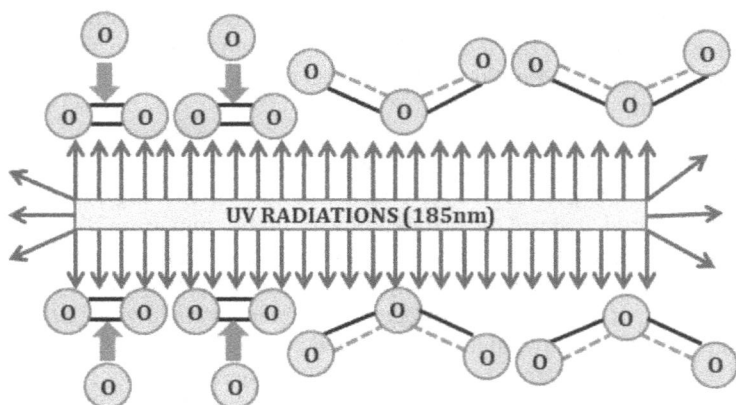

Figure 4.2 Assembly for photochemical or ultraviolet radiation–induced ozone production.

these photochemical generators end up limited in functionality from wastewater treatment for utilities only. Although it is possible to manufacture ozone for fumigation by using UV-based photochemical generators. This system of ozone generation ensures negligible production of nitrous oxide and is less impacted by humidity. Ultraviolet radiation with a wavelength of 254 nm also offers efficient disinfection of the air feed into the unit (Figure 4.2).

4.3.3 Electrolysis-Induced Ozone Production

When an aqueous system (at ambient temperature) containing chemical entities with phosphate groups is treated with radically elevated current intensities, it induces a swift lysis/breakdown (electrolysis) of phosphate groups. The electrolysis of these groups results in the production of ozone and oxygen molecules in the anodic gas. Electrolysis of 68 weight % sulfuric acid can produce up to 18–25 weight % ozone when a well-cooled cell is employed. Even while the lysis of water molecules by an electric current also produces ozone in notable quantities, the production efficiency is lower and the cost of operation is significantly higher in this process, in contrast to the plasma/corona discharge method. Nevertheless, small-scale electrolysis-based ozone-generating assemblies are used industrially for processing superior-grade water that is intended for the pharmaceutical and electronic industries (Figure 4.3).

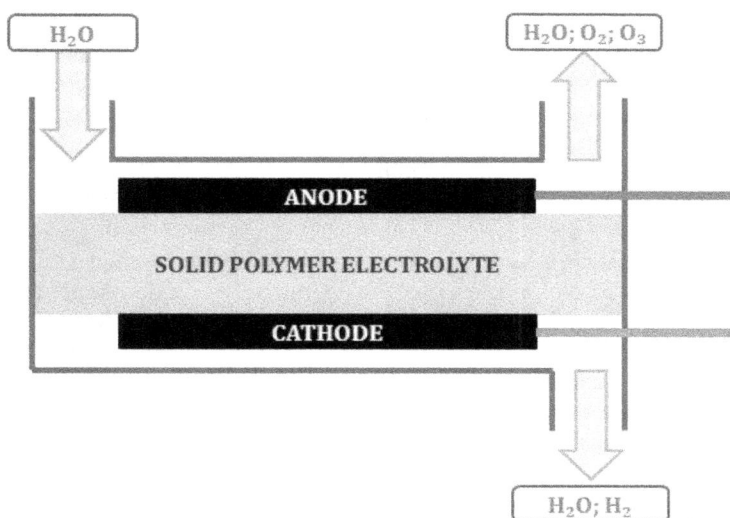

Figure 4.3 Assembly for electrolysis-induced ozone production.

4.4 Approaches for Ozone Utilization

There are many modes of utilization of ozone intended for microbial inactivation in various food products. The mode of utilization may be carefully chosen in accordance with the nature of the food matrix and the need behind ozone utilization. A broad spectrum of ozone generation systems with varying capacities are available, primarily differentiating in the nature of the feed/supply gas accessible for the synthesis of ozone. The purity and strength of produced ozone vary according to the system (aqueous and gaseous) employed.

4.4.1 Utilization in an Aqueous System

The half-life of ozone in a gaseous state is considerably higher than that of water-solubilized ozone, as it degrades very slowly. The presence of impurities such as organic compounds in an aqueous system prompts a simulated degradation of solubilized ozone. The process of ozone degradation is quite swift, particularly in an impure aqueous system. Ozone's low solubility in aqueous media necessitates adequate distribution of ozone (gas) in the form of tiny bubbles. This may be done with ease using various positive pressure ozone contractors, such as bubble diffusers and bubble columns, mechanically agitated vessels, turbine mixers, tube reactors, in-line static mixers, as well as negative pressure reactors (venturi) and injectors. In the case of positive pressure-assisted agitation, the gaseous bubbles are efficiently clipped and simultaneously blended with the aqueous system. This approach not only reduces the thickness of liquid bubble film, it also enhances the interfacial area and contact time considerably. Faster ozone mass transfer quotients result in superior rates of oxidation-induced microbial inactivation. However, organic entities that demonstrate sluggish oxidation behavior remain unaltered even at considerably augmented ozone mass transfer rates. To tackle these scenarios, advanced oxidation approaches are taken into consideration. The ozone concentration can be efficiently reduced to a tolerable limit for acceptable flux to the local environment by various thermal and/or catalytic means. It is occasionally possible to remediate effluent water that contains low ozone concentrations by passing it over wet granular activated carbon beds.

4.4.2 Utilization in a Gaseous System

In a gaseous system, the degree and speed of molecular interactions between ozone and generic air contaminants is significantly less and is slower in comparison to an aqueous system. Unlike in aqueous systems, regulating pH is not possible in the case of gaseous systems; but one practical solution is to increase the assembly's relative humidity. One of the most convenient ways to achieve disinfection through gaseous ozone is by pumping the gas from the generator assembly into the space of interest. The gaseous ozone molecules react promptly when they come into contact with contaminated entities such as volatile odorous molecules or airborne microbial agents. When the target air-based contaminants are more resilient and can only be slowly affected by ozone, an ideal approach to disinfection is mixing the tainted air with desired levels of ozone for the time needed to obliterate (or inactivate) contaminants.

4.5 Factors Affecting the Effectiveness of Ozone Processing

There are numerous fundamental factors associated with ozone processing that have a noteworthy impact on its effectiveness. These factors not only influence the reactivity, stability, and solubility of molecular ozone in aqueous systems, they also have substantial impact on its potency for microbial inactivation. Depending upon the nature of their relationship with respect to the aqueous disinfecting system, they are broadly categorized into two categories: external and internal factors.

1. *Internal factors* take into account the innate traits of the food matrix that have a significant effect on the efficacy of ozone, for example, pH, organic entities, residual ozone, and ozone demand of the system.
2. *External factors* take into account the native traits of the processing system that have a significant effect on the effectiveness of ozone, for instance, flow rate, ozone concentration, temperature, and relative humidity.

4.5.1 Internal Factors

4.5.1.1 pH

According to Patil *et al.* (2010), the impact of pH on ozone-assisted microbial disinfection or inactivation can be largely credited to the fact that the breakdown rate of ozone fluctuates considerably with changes in pH conditions. Kim *et al.* (2003) postulated that alkaline pH conditions result in prompt degradation of molecular ozone owing to the synthesis of hydroxyl radicals (OH⁻). As a result, the efficiency of molecular ozone–induced microbial inactivation changes with the changing pH of the aqueous medium during the course of the disinfection process. Wysok *et al.* (2006) showed that an aqueous disinfecting medium with a pH value ≥8 resulted in degradation of approximately half of the sparged ozone into a broad spectrum of intermediary oxygen species within 10 minutes. While investigating the ozone-assisted inactivation of *E. coli* Patil *et al.* (2010), found that the rate and efficiency of bacterial inactivation demonstrated a noteworthy positive correlation with the acidic pH, especially pH less than 5. Four minutes resulted in a 5-log reduction at a pH value of 5.

4.5.1.2 Organic Matter

A wide range of characteristics, such as several dispersed or suspended solids, organic and inorganic substances, solids, semi-solids, or liquid waste particles, jointly dictates the ozone demand of an aqueous system. Out of all these, the amount of organic substances significantly influences the microbial disinfection activity of the ozone treatment, as it drops with an increase in organic residue. The major reason behind the diminishing microbial inactivation potential of molecular ozone in the aqueous medium is its consumption by dissolved organic compounds. The organic moieties tend to induce a substantial degradation of molecular ozone into a broad spectrum of by-products with no to slight disinfection activity, thereby lowering the overall effectiveness of disinfection. In 2004, Williams *et al.* assessed the rate and effectiveness of ozone treatment for *E. coli* inactivation in citrus juice. They investigated the impact of organic residues such as ascorbic acid and sugar moieties on the efficacy of ozonation. They postulated that as the concentration of organic constituents increases, the degree of *E. coli* inactivation decreases. The presence of various organic entities, such as amines, proteins, lipids, and carbohydrates, in wastewater considerably raises the concentration of solubilized carbon (organics). Disinfectants such as ozone that specifically work by virtue of oxidation tend to lose their disinfecting potential owing to their interaction with organic entities. The breakdown reactions end up generating a range of by-products with lower to no disinfection strength.

4.5.1.3 Residual Ozone

"Residual ozone" refers to the quantifiable traces of ozone left in the treatment medium once it has been utilized for the disinfection of the target food commodity. As discussed earlier, the overall concentration of ozone sparged in an aqueous medium is the chief factor with an impact on the overall efficacy of the disinfection process. However, residual ozone is also a crucial component that exerts an equivalent influence on the efficacy of disinfection in opposing microbial inactivation. Ozone-demanding entities present in an aqueous medium together with the stability of the disinfectant under specific conditions of application prominently influence the potency and efficacy of ozone-based microbial inactivation. For ozone-based disinfection, it is extremely crucial to report certain parameters during disinfection treatments, such as decay rate, available concentration, and applied dosage; otherwise, there is a high possibility of overestimation of the actual effective dose.

4.5.1.4 Ozone Demands in Media

Gardoni *et al.* (2012) compared the ozone demands of different treatment media with pure water and found that pure water had the minimum ozone demand. Organic and inorganic contaminations in water interact with sparged ozone molecules and create the demand for ozone. Humic compounds, glyoxylic acid, and formic acid, for example, stimulate the breakdown of ozone and contribute significantly to the increased demand for ozone. Different grades of purity of water systems impact the solubilization of ozone; for instance, Gardoni *et al.* (2012) found that the rate of ozone solubilization in deionized and distilled aqueous systems is quite rapid. Upon comparison, it was confirmed that ozone's solubility in distilled and deionized water samples was double than that of tap water. On the other hand, a detailed comparison between the solubilization of ozone in distilled and deionized water systems demonstrated that the rate of solubilization was rapid in the latter system. Furthermore, food matrices with naturally occurring antioxidants tend to create an add-on demand for ozone by scavenging the active radicals produced by ozone breakdown. A broad spectrum of sweeteners, chelators, emulsifiers, sequestering, and preservative agents can influence the stability of ozone contingent on their intrinsic properties.

4.5.2 External Factors

4.5.2.1 Flow Rate

The conclusive dimensions of the ozone bubble in the treatment matrix are positively reliant on the flow rate of ozone sparging employed for the production of the disinfection medium. By manipulating the flow rate of ozone sparging, we can very efficiently modify the bubble sizes of molecular ozone. The resultant dimensions of ozone bubbles in the treatment medium categorically influence the two most crucial physicochemical characteristics of disinfecting molecular ozone: efficiency of microbial inactivation and rate of ozone solubilization. In 2006, Akbas and Ozdemir found a vital correlation between the sizes of ozone bubbles (mass transfer of ozone) with the effectiveness of microbial inactivation when other variables were kept constant. They proposed that a noteworthy reduction of sparged ozone bubbles not only significantly enhances the mass transfer of ozone, but it also simultaneously improves the efficacy of microbial inactivation.

The small size of molecular ozone bubbles categorically demonstrates a high interfacial cross-sectional area that is available for rapid interaction with microorganisms, as well as ensuring significant mass transfer. Gardoni *et al.* (2012) proposed that a 10-fold reduction in ozone bubble size consequently augments its interfacial cross-sectional area by a factor of 32. Detailed studies of microbial inactivation kinetics suggest that because of the presence of surface-active agents on their cellular membranes, free-flowing and dispersed bacteria have a distinctive propensity to travel in the direction of ozone bubbles. As a result of this tendency, they are preferentially incapacitated by relatively superior concentrations of ozone precisely at the interface between the gas and liquid phases of the bubble.

4.5.2.2 Concentration

The resultant effective concentration of molecular ozone (available for interaction) in the disinfecting medium is an additional component governing the overall effectiveness of the disinfection process. Sparging of ozone in excessive amounts causes the concentration to go above the saturation levels, thereby rendering the subsequent influx of the gas to the assembly chemically unproductive, followed by extended durations to attain an equivalent degree of microbial inactivation.

4.5.2.3 Temperature

For the first time, Rice (1986) and Bablon *et al.* (1991) conducted a series of experiments where they examined the facilitation of ozone solubility with the lowering of the temperature of the aqueous medium. They proposed that at a temperature condition of 0–30°C, the solubility of molecular ozone in aqueous medium is 13 times more than the solubility of oxygen with a corresponding decline of solubilization temperature. An extensive increase in the degradation or breakdown rate of ozone owing to the remarkably low stability and solubility of ozone at high temperature conditions is responsible for its ineffectiveness. Quite the opposite is the enhanced reactivity of the residual ozone at a high temperature. The mass transfer of molecular ozone is substantially influenced by the temperature of the water phase. The effectiveness and efficiency of ozone-induced microbial inactivation are in coherence with the decreasing temperature. The processing conditions maintained while carrying out disinfection of produce have a great collective influence over the solubility, stability, reactivity, and efficacy of ozone.

4.5.2.4. Relative Humidity

Many researchers have investigated the significance of high humidity in ozone-induced inactivation of microorganisms. Galdeano *et al.* (2018) found that ensuring effective microbial inactivation prompted by ozone gas requires the utilization of high humidity in the disinfection process. They demonstrated that highly humid atmospheric conditions ensure an adequate hydration of dehydrated microorganisms, thereby rendering them vulnerable to ozone. A relative humidity of approximately 90–95% is imperative to ensure adequate microbial inactivation by molecular ozone in a gaseous state. In contrast to this essential requirement of sustaining high relative humidity to facilitate an effective microbial inactivation, ozone itself is degraded promptly at high humidity in comparison to low humidity conditions. Gardoni *et al.* (2012) scrutinized the potency of low concentrations in carrying out ozone-based disinfection of airborne microorganisms at variable humidity conditions and long exposure time. They witnessed a remarkable decline in the microbicidal supremacy of ozone at RH < 45%, designated insignificant. On the contrary, the microbicidal activity of ozone increased greatly at high humidities, even at nominal to marginal concentrations of ozone (<0.1 mg L^{-1}). As a result, ozone is established as an efficient sanitizing agent exclusively active against hydrated microbes. It instigates their rapid killing in a relatively high-humidity environment.

4.6 Structure for Ozone-Instigated Microbial Inactivation

A broad range of process dynamics like early or preliminary microbial population of the raw commodity, indecorous and inappropriate handling and storage of the products (specially fruits and vegetables), usage of impure or low quality grade water for washing the products, inadequate processing machinery, inefficient transportation amenities, gross cross contamination of the fresh products from other commodities etc. considerably subsidizes the biological hazards (microorganisms) accompanied with different food commodities. As already discussed, ozone is a robust and resilient oxidizing agent, and its powerful oxidizing effect enables it to carry out the advanced oxidation of imperative cell constituents, thereby causing the obliteration of microbes. These potent oxidizing alterations in the cellular components instigate permanent injuries to the phospholipid bilayer, carrier proteins of the cell membranes, genetic material (DNA and RNA), and vital cell organelles such as mitochondria and endoplasmic reticulum. These modifications results in non-rectifiable impairments of the microbial cells that hamper their normal functioning and survival. Efficiently hydrated microbes are more radically exterminated by ozone than the dried ones; that is, the RH of the processing atmosphere plays a crucial role in achieving oxidization by ozone. The chief target of ozone oxidation is the microorganism's cell surface. Microbial cells are exterminated by structural interruption of their cell membranes, followed by the disintegration and lysis of the cell.

The following fundamental mechanisms are acknowledged for the targeted inactivation of microbes in food products:

Mechanism 1: The microorganism cells, containing proteins in the form of intercellular enzymes, peptides and vital proteins and carrier proteins present to facilitate intercellular transportation are extensively oxidized by the molecular ozone and subsequently fragmented into smaller peptides.

Mechanism 2: Considerable oxidation of polyunsaturated fatty acids is present as an integral constituent of the lipopolysaccharide layer or cell membrane of Gram-negative bacteria. Oxidized to acid peroxides by ozone and upon further oxidation, it forms prominent pores in the cell membrane that eventually initiate leakage of the vital cellular components. In addition to this, sometimes the genetic vitals of the cell are hampered, which eventually results in cell death.

Mechanism 3: Molecular ozone shows significant potency to interact with the microbial cell dehydrogenases and genetic material of the cell, that is, DNA and RNA.

Many researchers have shown or postulated that molecular ozone is the prime microbial inactivation agent (direct inactivation). On the contrary, a number of authors have also credited the bactericidal potential of the molecular ozone to the reactivity of ozone breakdown products, for instance, OH, O_2-, and HO_3, as the fundamental mediators of microbial inactivation. Molecular ozone along with its highly reactive breakdown species performs a vital role in the obliteration of microorganisms, but there is no clear consensus on which particular component is most significant in ensuring microbial inactivation. There are two chief interactions transpiring on the prevalence of molecular ozone over the microorganism's membrane by means of indirect and direct interactions. For instance, direct and/or indirect pathways could be responsible for the oxidization of unsaturated fatty acids along with sulfhydryl moieties. Both of these are vital for conserving the structural cellular integrity and ensuring proper functioning of enzymes.

4.6.1 Direct Contact–Induced Microbial Inactivation

The principal mechanism for the obliteration of microbes is direct interaction with molecular ozone by virtue of complete oxidation of the prime target sites in the microorganism's cell membrane prompted by molecular ozone. It is convincible that direct interaction–induced microbial inactivation relative to the outcome of an indirect reaction may vary significantly with respect to different microorganisms.

By virtue of a broad range of factors, for instance, distinctive structure, typical spatial arrangement of molecules, resonating hybrids of molecules, and distinguishing mechanisms of interaction with organic and inorganic species individually, ozone is a molecule that very proficiently acts as an electrophile as well as a nucleophile. An electrophile is a reactive species that formulates chemical bonds by accepting an electron pair with nucleophiles. On the other hand, a nucleophile is molecular species that establishes chemical linkages by donating an electron pair to electrophiles. On the whole, an aqueous matrix has substantially high quantities of organic compounds as impurities and contaminants, thereby attaining a remarkably high electron density and unquestionably triggering instant electrophilic reactions. Correspondingly, electrophilic interactions ensue swiftly in solutions with high quantities of aromatic compounds. On the contrary, if the disinfection matrix is rich in compounds containing carbon atoms, that is, electron-withdrawing groups such as $-COOH$ and $-NO_2$, and there is an overall deficiency of electrons, nucleophilic interactions are prompted. Thorough investigations of reaction kinetics have shown that the overall speeds of these interactions are lower than electrophilic reactions. In addition, it's imperative to note the influence of the pH of the aqueous system on the breakdown of sparged ozone; a pH value more than 7 (alkaline) prompts a rapid breakdown of the ozone molecules, whereas intensely acidic conditions (pH values of less than 3) do not affect ozone decomposition significantly.

4.6.2 Indirect Contact–Induced Microbial Inactivation

Indirect interactions taking into account free radical species were found to be responsible for the inactivation of some specific classes of microorganisms. There are three distinct stages of ozone breakdown: initiation, propagation, and termination. During the course of initiation, highly reactive radical species such as superoxide radical ions and hydroperoxide radicals are engendered, which collaboratively result in the generation of exceptionally reactive hydroxyl radicals. These hydroxyl radicals are one of the prime factors responsible for the breakdown of molecular ozone. After the initiation stage comes the propagation stage that takes into account repetitive regeneration of the hydroperoxide and superoxide radicals by virtue of interactions including the contribution of a broad spectrum of promoters, for instance, formic acid, glyoxylic acid, primary alcohols, and aryl groups, in contrast to the termination stage, where the consumption of hydroxyl radicals occurs via interaction with ions like bicarbonate, carbonate, tertiary alcohols, and alkyl groups deprived of superoxide radical ion regeneration (Staehelin and Hoignr, 1985; Khadre and Yousef, 2001).

The eukaryotic and prokaryotic cells of microorganisms usually show the presence of bicarbonate ions, which probably perform as scavengers of highly reactive free radical species or else are responsible for the obliteration of the microbes. Moreover, characteristics prompting and enhancing the breakdown of molecular ozone in the aqueous system can categorically end in swift breakdown, thereby increasing the overall ozone demand and projecting it as a necessity of amplified ozone concentration in order to accomplish the desired level of microbial disinfection. The subsequent disruption or breakdown of microbial cell membrane concurrent with the ozone is a faster obliteration mechanism than those of other disinfectants. That in a way makes the permeation of disinfecting agent into the cell envelope which is an imperative feature to ensure its effective and efficient inactivation. In general, with regard to the range of disinfecting action, every individual microorganism shows a distinguishing inherent vulnerability towards ozone. Molds are least affected by ozone, followed by yeast and then bacteria. Furthermore, bacteria spores demonstrate substantially high resistance compared to vegetative cells, and Gram-positive bacterial cells are more vulnerable than Gram-negative cells. Due to the mechanism of ozone action, which destroys the microorganism through cell lysis, the development of resistance to ozone disinfection is not found.

4.7 Target Locations for Ozone Action

Various investigations have established that ozone molecules strike several cellular components simultaneously. These take into account the structural building blocks of the cell membrane (phospholipids and proteins), enzymes that are vital for the normal functioning of the cell, genetic material such as DNA and RNA, and the sturdy coatings of spores and capsids in the case of viruses. This much diversification in the target sites means the obliteration of microorganisms via ozone is a complex process. The chief cellular components with crucial interactions with ozone are discussed in the following.

4.7.1 Cell Membrane

As shown in the classic fluid mosaic model, the structural building blocks of the cell membrane are a phospholipid by layer with intermediary carrier protein channels governing the overall permeability of the microbial cell. So ozone, owing to its remarkable oxidizing potential, tends to oxidize these vital constituents along with membrane-bound enzymes, glycoproteins, and glycolipids, thereby hampering the structural and functional viability of the cell, for instance, prompting pore formation and outflow of cellular innards and eventual cell lysis. Dave (1999), with the help of electron micrographs in treatment of *Salmonella enteritidis* with aqueous ozone, demonstrated that as the unsaturation present in the phospholipids and sulfhydryl groups of the enzymes is oxidized by the ozone, the native pursuit for cellular metabolism, permeability and sustainability is completely destroyed, and a swift cellular death follows.

4.7.2 Bacterial Spore Coats

Galdeano *et al.* (2018), while assessing the contribution of spore coat proteins to ozone inactivation, witnessed that spore coat protein functions as a primary protective, strong, and sturdy shield for spores; decreases their susceptibility; and maintains homeostasis in hostile conditions. They reported that in case of *Bacillus cereus*, the bacterial spores with the removal of spore coat proteins were found to be significantly vulnerable to ozone treatment and were inactivated rapidly in comparison to the native undamaged spores.

Many researchers have conducted studies to determine the factors influencing the strength and integrity of this primary protective barrier that makes spores relatively resistant to any microbicidal treatment. Galdeano *et al.* (2018) further elucidated that since the spore coat constitutes more than 50% of the spores by volume, it is difficult for any disinfecting agent to permeate this barrier and bring about the desired inactivation. In the case of ozone-prompted inactivation of *Bacillus* spores, the permeation of ozone through the spore coat was relatively tough, and during the course of its propagation into the spore, there was a gradual decline in the powerful oxidizing action of ozone. In 2004, Scott and Lesher proposed that ozone-mediated inactivation of bacterial spores wasn't because of its detrimental effect on DNA, but rather the damage is because of the destruction of ability of the spores to germinate. They anticipated the loss of germination power could be because of non-rectifiable injuries in the innermost membranes of the spores.

4.7.3 Cellular Enzymes

Ozone-induced enzyme disruption is a noteworthy mechanism that inactivates microorganisms. In 1965, Sykes compared the disinfection function of ozone with that of chlorine for the first time and observed that chlorine demonstrated a selective disruption of enzymes, whereas ozone was a potent oxidant for the entire protoplast. The inactivation of cellular enzymes is principally because of their structural disruption by ozone-induced oxidation of cysteine residues (Choi, 2005).

4.7.4 Genetic Material

The obliteration of microorganisms is also credited to the damage of the nucleic material, that is, DNA and RNA, upon reaction in an aqueous system with molecular ozone. In nucleic material, various purine and pyrimidine residues demonstrate variable susceptibility towards ozone; for instance, uracil is most stable, followed by cytosine and thymine. McKenzie *et al.* (1997) proposed that ozone uncoils the native functional structure of DNA and subsequently destroys its ability to transform. It instigates multiple breaks in the structure, resulting in a decrease in DNA transcription activity. It also prompts genetic mutations, but in contrast to other known mutagenic agents, the potential of ozone was significantly inferior.

4.8 Synergy of Ozone with Different Sanitizing Agents

With the advent of ozone progressive oxidation, there are recent instances described of the combination of ozone with either hydrogen peroxide or ultraviolet radiation, techniques that are suggested to enhance the formation of hydroxyl free radicals with the stated objective of intensifying the microbial inactivation potential of the resultant combination above that of ozone itself. Advanced oxidation process techniques are designed to promote the formation of hydroxyl free radicals, resulting in increased microbial inactivation potential above that of ozone itself. Effective ozonation procedures are also being developed through improved delivery systems in order to overcome the physical barriers that diminish the efficacy of sanitization of products and to maximize the biocidal action of ozone. Some recent reports indicate what appears to be a synergy or an increased potential of microbial inactivation by applying these combinations to certain foodstuffs over what is obtained by applying ozone or UV radiation alone.

4.8.1 Ozone and Hydrogen Peroxide

With the combination of ozone and hydrogen peroxide, both are powerful oxidants in aqueous solutions that eventually destroy each other and mutually give rise to hydroxyl free radicals. It is a customary technique to add the requisite amount of hydrogen peroxide to a solution and then pass that solution through an ozone-contacting apparatus. Ozone reacts instantaneously with hydrogen peroxide present in the solution. Care should be taken that the amount of ozone dosed in the contactor is greater than the amount of peroxide initially supplied to the solution. In advanced oxidation processes, each pollutant that needs to be destroyed requires a specific weight ratio of peroxide to ozone to show the optimum oxidative performance.

It is advisable to first determine which polluting constituents requiring treatment are present in the aqueous system, and then the optimum range of peroxide to ozone weight ratio required for their destruction is determined experimentally. If excess hydrogen peroxide is present over the amount of ozone added, at least some of the advantages of advanced oxidation are lessened. It also means that there will be no molecular ozone present at any time during ozone contact for microbial disinfection.

4.8.2 Ozone and UV Radiation

It is usual to place a UV bulb (or multiple bulbs) in the ozone-contacting chamber with this combination of agents. As water propagates through the chamber, first the UV bulb(s) is (are) turned on and ozone is added. On the condition that the amount of UV radiation dosed is in excess of the amount of ozone present, all ozone will be converted instantaneously to degradation products, ending swiftly as hydroxyl free radicals. Several reports have been made of an apparent increase in antimicrobial activity in some food applications when ozone is combined in water with either peroxide or UV radiation.

4.9 Functionality of Ozone Processing in Food Industries

In the current scenario, the global markets are full of informed customers with new preferences, be it in terms of processing, composition, shelf life, or health aspect of food products. Customers are gradually showing a preference towards food commodities that are minimally processed and free from chemical preservatives, and these requirements are strengthening progressively. Apart from these changes in consumer thought processes, the other factors that play a crucial rule in the transition towards minimally processed food products, innovative processing, and preservation techniques are contemporaneous outbreaks of foodborne pathogens and identification of new foodborne pathogens. Novel food-processing and preservation techniques very proficiently guarantee food safety by curtailing the incidences of disease-causing microbes along with conserving the expected quality characteristics by countering spoilage-causing microbes in processed goods. Owing to all of these reasons, the use of ozone, along with additions such as UV radiation or hydrogen peroxide, is an accepted practice these days that results in superior end product qualities with remarkable cost savings. In 1982, the United States Food and Drug Administration (USDA) granted the use of ozone in bottled water "generally recognized as safe" (GRAS) status for use.

4.9.1 Raw Poultry and Meats

The shelf life of raw poultry and meats is governed by the load and diversity of the microbial population. The microorganisms that inhabit the gastrointestinal tract administer the variety and number of microorganisms existing on raw poultry and meats. Furthermore, in addition to numerous spoilage microorganisms, these products sporadically carry pathogenic microorganisms such as *Campylobacter* spp., *Salmonella* spp., *L. monocytogenes*, and pathogenic *E. coli*. Several stages in conjunction with the food chain, including slaughtering, handling, storage, and distribution, enable contamination of food products with these microorganisms. At present the employment of sanitizers on carcasses and cut meat is limited. Ozone was permitted by the U.S. Department of Agriculture for reconditioning recycled poultry chiller water in 1997. Beef and beef brisket fat decontamination by ozone was investigated by a number of researchers; results were variable. Mitsuda *et al.* (1990) observed that the use of gaseous ozone minimized or prevented growth of microorganisms on the meat surface.

Several investigators have demonstrated microbial inactivation efficiency and safety of ozone for use in washing poultry carcasses (Izat *et al.*,

1990; Jindal *et al.*, 1995), reconditioning poultry chiller water (Waldroup *et al.*, 1993; Diaz and Law, 1999), and sanitizing hatchery equipment (Whistler and Sheldon, 1989). Utilization of a synergistic combination, that is, ozone with a suitable adjunct (such as hydrogen peroxide and UV radiation), in the form of a progressive oxidation process positively enhanced the efficiency of ozone as an antimicrobial control agent in poultry chiller water (Diaz and Law, 1999). The presence of *Salmonella enterica* ser. *Enteritidis* in shell eggs has serious public health implications. In order to control or eliminate this particular pathogen in eggs, a number of thermal and chemical treatments have been developed.

Yousef and Rodriguez-Romo (2001) used ozone at low temperatures and mild pressure for cold sanitization of shell eggs. Such a treatment could be used industrially to produce "cold-sanitized" eggs. Cox *et al.* (1995) patented a procedure that uses vacuum, heat, and ozone treatment to eliminate *Salmonella* spp. from the shell eggs and defined it as "hyper pasteurization". This method includes heating shell eggs at higher than 54.4°C for longer than 15 minutes, with successive application of ozone. According to this report, the combined treatment extended the shelf life and reduced the microbial load of shell eggs.

4.9.2 Fruits and Vegetables

There is a broad spectrum of inappropriate food production practices that play a vital role in causing substantial pre-harvest contamination of produce. These practices are utilization of manure fertilizer, heavily contaminated irrigation water, and poor hygiene of farm and industry people dealing with horticultural produce. Similarly, after the harvest of horticultural crops, factors such as inappropriate handling of produce, followed by inadequate storage, usage of contaminated water for removing field heat and washing, unsuitable transportation amenities, and chances of cross-contamination from other produce further increase the occurrence of biological hazards. Out of all these practices, those of extreme concern are the microbiological quality of the aqueous media employed for washing produce as well as the chances of cross-contamination from other commodities. Processing operations such as peeling, coring, dicing, slicing, shredding, or cubing play a vital role in disrupting the integrity of the surface. As a result of this surface integrity disruption, the produce is significantly susceptible to rapid microbial contamination. Remarkable enhancement of the microbial safety of freshly produced horticultural crops is attainable by the employment of a suitable ozone treatment. Ozone-assisted disinfection can accomplish the mentioned goals and also help better the recyclability of effluent generated in the fruit and vegetable processing industry.

There are many validated and successful investigations carried out to assess the efficiency of ozone as a potential disinfectant for fruits and vegetables. Liu *et al.* (2021) investigated efficient disinfection and utilized aqueous ozone with a concentration of 1.4 mg L^{-1} at variable disinfection treatment times of one, five, and ten minutes. Apart from inactivating the microbes, ozone also demonstrates certain additional benefits. It reduced respiration, thereby lowering ethylene synthesis, and enhanced the quality of freshly shredded cabbage compared to a control. Ozonated water with a treatment exposure of five minutes was found to inhibit the growth of aerobic and coliform bacteria as well as yeasts. The microbicidal activity of ozone was further enhanced with an exposure time of ten minutes. Karaca and Velioglu (2020) compared the efficiency of aqueous ozone with a concentration of 12.0 mg L^{-1} against chlorine at a concentration of 100 mg L^{-1} and an exposure time of five minutes for the disinfection of parsley leaves. They found that aqueous ozone didn't adversely impact the chlorophyll, vitamin C, total phenolics, or antioxidant potential of the freshly cut leaves. They also found a remarkable decline in *E. coli* counts in the samples treated with ozone. On the other hand, ozone was more effective in inhibiting the growth of *L. innocua* in comparison to chlorinated water; hence, it could be a substitute for chlorine. Chen *et al.* (2020) utilized aqueous ozone with a concentration of 1.4 mg L^{-1} for the disinfection of onions at variable disinfection treatment times of one, three, and five minutes. They observed that an exposure time of one minute remarkably lowered the weight loss of freshly cut onions, even during the course of significantly long storage of 8 to 14 days. All three treatment times were found to lower the rate of respiration and subsequent softening of the onions during storage. Apart from ensuring significant disinfection and vital changes in the quality of the onions during storage, ozone also caused a noteworthy decline in pesticide residues from the most common pesticides: dimethyl dichlorovinyl phosphate, cypermethrin, chlorpyrifos, methomyl and omethoate. Ozone demonstrated valuable bactericidal effects against coliforms, yeast, and aerobic bacteria. Wang *et al.* (2019) investigated and compared the proficiency of aqueous ozone for the disinfection of lettuce as a sole disinfecting agent (1 mg L^{-1} for 30 seconds) or in combination with lactic acid (2 mg L^{-1} together for 30 seconds with 1% lactic acid for 90 seconds). The subsequent assessment of quality for the lettuce in the characteristics of color, texture, exudate seepage, total phenolics, and weight loss found that ozone demonstrate a synergistic behavior with lactic acid. It not only ensures better inactivation of microbes such as *E. coli* O157:H7 and aerobic, mesophilic, and psychrophilic molds and yeasts, but it also maintains the integrity of quality characteristics during the storage of the produce.

4.9.3 Fish and Seafood

Fish and seafood are extremely perishable food commodities, which restricts their consumption in a reasonably fresh state to the immediate vicinity of where they are caught. The spoilage of these food products is specifically associated with the bacteria that degrade fish constituents, particularly non-protein nitrogenous compounds. The nature of fish and seafood species, handling, and storage conditions are the most vital features that affect their spoilage. Different food processing and preservation techniques have been applied in order to reduce the perishability of these commodities and extend the shelf life. Since the 1920s, scientists have attempted to employ the powerful disinfection properties of ozone to slow down the spoilage and improve the safety of fishery products. Mosayebi *et al.* (2010) found that an appropriate treatment by ozone prior to any type of storage improves the microbiological and biochemical qualities of fish and seafood and consequently prolongs their shelf life. The shelf life of these foods can be considerably extended by implementing an efficient combination of ozone treatment and cold storage (at 4°C). Ozone treatment slows down bacterial growth significantly, resulting in lower counts of bacteria. Subsequently, ozone has no adverse impact on the biochemical properties of the product such as humidity, protein, and free fat. Ozone leaves no residue on the products and creates no changes in color and flavor.

Ozone was tested for decontaminating shrimp (DeWitt *et al.*, 1984); mussels (Abad *et al.*, 1997); and various fish such as jack mackerel (Haraguchi *et al.*, 1969), sockeye salmon (Lee and Kramer, 1984), Japanese flounder, and rockfish (Mimura *et al.*, 1998). The absence of adverse sensory effects and harmful oxidation by-products confirms the desirability of ozone use in processing fish products for human consumption. Ozone treatment of shrimp meat extract, however, was ineffective against microorganisms in the product (Chen *et al.*, 1992). Ozone may have reacted with ozone-demanding substances in the extract instead of microbial cells. Abad *et al.* (1997) developed the use of ozone for shell fish depuration from the laboratory stage to full commercial installations in southern Europe. Depuration is a process whereby shellfish, freshly harvested from their natural environments, are placed for several days in storage chambers through which clean, pathogen-free water is passed. Ozonated water was used for depuration. Over several days, the mollusks cleanse themselves by passing disinfected water through their systems, thus eliminating pathogenic microorganisms imbibed from their natural environments.

4.9.4 Milk and Milk Products

Removal of milk residues and biofilm-forming bacteria from stainless steel surfaces by pre-rinsing with warm water is normally the first step in cleaning dairy processing equipment in order to remove the bulk of milk residue (a.k.a. dairy soil). Guzel-Seydim *et al.* (2000) quantified and showed the effectiveness of warm water (40°C) and ozonated cold water (10°C) as a pre-rinse for removing dairy soil from stainless steel plates. Scanning electron micrographs showed that the metal surfaces were cleaned more efficiently by ozonation than by a 40°C warm water treatment. According to the results of chemical oxygen demand (COD) measurements, ozonated water removed 84% of milk residue from plates, whereas the non-ozonated warm water treatment removed only 51% of dairy soil materials. Similarly, Fukuzaki (2006) and Jurado-Alameda *et al.* (2014) studied the suitability of ozone for removal of heat-denatured whey proteins from stainless steel surfaces. Both aqueous and gaseous ozonation facilitated whey protein desorption. Micro-organisms adhering to milk contact surfaces are hard to destroy and may cause deterioration in the microbiological quality of milk and dairy foods. Ozonation is a possible alternative to the chlorine-based sanitizers widely used in the dairy industry (Guzel-Seydim *et al.*, 2004).

In order to make raw milk safe for human consumption, it is traditionally treated with diverse thermal treatments. Heating, however, may negatively influence both the nutritional value and sensory properties of milk. For this reason, Sander (1985) patented a method for the mild ozone treatment of liquids, including milk and fluid dairy foods, thereby minimizing their possible quality deterioration. Rojek *et al.* (1995) used pressurized ozone (5–35 mg/L for 5–25 min) to preserve skim milk by decreasing its microbial populations. The treatment was shown to reduce the number of psychrotrophs by more than 99%. Ozonation alone is not capable of killing a sufficiently high percentage of the microbial load of raw milk. A gentle process involving pre-ozonation followed by a conventional pasteurization step has been developed by a Swedish company Pastair. The treatment is claimed to result in commercial fluid milks with an extended shelf life without causing excessive lipid or protein oxidation in the final products. Ozone treatments may also influence the chemical, physical, functional, and organoleptic properties of dried milk products. It has been found that ozonation substantially enhances the foaming capacity and foam stability of proteins; however, both the solubility of whey proteins and the emulsion stability are lowered. It is noteworthy that gaseous ozone treatments decreased the solubility of protein samples to a greater degree than did aqueous ozonation. Similar findings were reported by Segat *et al.* (2014), who concluded that tailored whey proteins with specific functionality may be developed through ozone processing.

4.9.5 Food Grains and Their Products

The control of pest (insects and microorganisms, including mold, fungi, and bacteria) development in stored grains after harvest is essential, as it currently leads to substantial food grain yield loss around the globe. Storage grains are especially susceptible to a number of insects, such as *Tribolium*, *Sithophilus*, and moths, which cause extensive destruction and could potentially acquire resistance to the currently employed insecticides. Ozone that can be used in fumigation is an interesting substitute for chemicals for the control of insect development (Sharma *et al.*, 2021). Kells *et al.* (2001) evaluated the efficiency of ozone fumigation in a corn grain mass against adult insects such as the red flour beetle (*Tribolium castaneum*), maize weevil (*Sithophilus zeamais*), and larvae from the Indian meal moth (*Plodia interpunctella*). Results demonstrated a significant insect mortality increase when the insect species in grain samples were treated with 50 ppm for three days. Due to its inactivating action on fungi, ozone can also be considered to help reduce mycotoxin accumulation during grain storage. Furthermore, its oxidant properties could be used for mycotoxin degradation and detoxification. Degradation of aflatoxin (McKenzie *et al.*, 1997), as well as trichothecenes, is started by the attack of a double bond with addition of two oxygen atoms, which further leads to the molecule breaking apart.

4.9.6 Sanitation of Other Materials

Ozone is a potent oxidant and ensures the effective obliteration of microorganisms via oxidation in comparison to the most commonly employed non-oxidant agent, chlorine, to avoid unacceptable changes in organoleptic characteristics and carcinogenic effects of food products. In comparison to chlorine in terms of effectiveness, ozone is 3000 times more potent and is GRAS. The possible future applications of ozone involve commercial refining of effluent generated from fish hatcheries, beverage producing plants, and wineries. The employment of ozone in the disinfection of food packaging material, processing lines, food contact surfaces, equipment, and vessels has produced extraordinary outcomes in the characteristics of regulating microbial growth and cost savings due to lower expenditure on handling and maintenance.

4.9.6.1 Sanitation of Food Packaging Material and Contact Surfaces

Even though the microbial count of materials intended for the packaging of food commodities is significantly less than that of the actual produce, there is a good probability of their survival through conventional disinfection treatments, which can later result in considerable hampering of food quality and safety. Protecting the quality and safety of food is one of the most important reasons why ozone aided sanitation procedures are so important to the food industry. A broad spectrum of disinfection approaches, thermal (heat), chemical (disinfectants such as H_2O_2 and chlorine), and radiation (UV), are currently employed individually or in combination to achieve the desired outcomes. Every strategy, meanwhile, has a variety of related benefits and drawbacks. For example, H_2O_2 disinfection has variable overall effectiveness in terms of results obtained, which makes the procedure unreliable. Second, unacceptable amounts of H_2O_2 breakdown products might remain even after disinfection and react with some of the polymers in the packaging material. Owing to these drawbacks, an efficient substitute has long been needed for the effectual sanitization of materials used for food packaging. Khadre and Yousef (2001) established the supremacy of molecular ozone in the sensitization of multi-laminated aseptic food packaging material and stainless steel to inactivate natural contaminants, bacterial biofilms (*P. fluorescens*), and dried films of *B. subtilis* spores. After being treated with 5.9 mg L^{-1} ozone in water for one minute, the multi-laminated packaging material—which included low levels of naturally occurring, primarily mesophilic contaminants—was rendered sterile. In order to decrease the microbial population effectively in biofilms, repeated exposure to ozone and agitation are required during the treatment.

4.9.6.2 Sanitation of Food Processing Lines

With the intention of safeguarding the effective disinfection of food processing lines such as batch tanks/silos, pipelines, pumps, and any surface that is in direct and immediate contact with a food commodity, water with sparged ozone is extensively used. The ability to recycle and reuse cleaning water is one of the most significant and practical benefits of using an ozone-assisted disinfection system in food processing companies. This type of sanitation system is known as a "clean in place" system. This could bring significant savings in the cost of disinfection systems.

4.10 Impact of Ozone Disinfection on Product Quality

The impact of ozone disinfection on various food quality characteristics, such as physicochemical, color, texture, and sensorial (aroma and taste) traits, varies significantly across commodities depending upon the native composition and nature of the constituents. The concentration of ozone and exposure time employed in the course of disinfection are also imperative factors that not only affect the overall microbial inactivation but also the characteristics of the food system under consideration. Superficial oxidation, loss of color due to structural degradation of pigments, generation of objectionable smells, and oxidative degradation of vitamins are a few of the unacceptable changes that occur in food products exposed to prolonged and excessive ozone treatment.

4.10.1 Physicochemical Characteristics of the Food System

The major constituents of the food matrix that is greatly influenced upon exposure to ozone are unsaturated constituents and/or compounds with antioxidant properties (Sharma *et al.*, 2022). Because of the remarkable oxidation potential of ozone, these molecules show considerable vulnerability towards ozone-induced structural deterioration. For instance, disinfection of freshly cut lettuce leaves with an aqueous system sparged with an ozone stream showed a significant influence on the final phenolic content of the produce. However, many contradictory studies have also been published on the influence of ozone on vitamin C.

4.10.2 Color Characteristics of the Food System

Gabler *et al.* (2010) examined the competence of ozone as a possible substitute for SO_2, the most widely employed commercial fumigant in post-harvest disinfection and management of post-harvest spoilage of table grapes during storage. They observed that the activity of grey mold (*Botrytis cinerea*, the most potent spoilage-causing factor in the case of table grapes) was efficiently controlled with an ozone treatment of up to 10 000 microliter/L for up to two hours.

4.10.3 Textural Characteristics of the Food System

One of the most dynamic and influential organoleptic characteristic of any food commodity is texture, be it the crunchiness of a freshly harvested fruit or vegetable, softness of freshly baked muffin, chewiness and juiciness of a grilled steak, creamy and smooth consistency of a soup, softness of cheese, or crispiness of breakfast cereals. These are few of the textural characteristics that are widely anticipated by consumers and play a significant role in their respective food preferences. For instance, in the case of a majority of food products, consumers often correlate textural characteristics with its freshness and goodness. There are various means by virtue of which the textural characteristics of food products are greatly modified, and they are usually categorized as non-enzymatic and enzymatic interactions. Disinfecting freshly harvested fruits and vegetables with an aqueous system sparged with ozone or after-processing storage of goods in an atmosphere flushed with ozone was shown to demonstrate characteristic modifications in the textural features of the respective commodities (Pandiselvam *et al.*, 2022). Out of all the fresh produce, green leafy vegetables are of major concern.

It has been shown by multiple researchers that the firmness, crispiness, and crunchiness of vegetables such as fresh coriander, spinach, cabbage, basil, parsley, and green leafy onions are greatly affected by ozone-assisted disinfection. For instance, An *et al.* (2007) reported considerable loss of firmness and crunchiness of the fresh coriander leaves when washed in an aqueous medium sparged with molecular ozone in comparison to that washed with normal water. On the one hand, the loss of firmness and other classical textural features during the course of post-harvest and/or post-processing storage (such as MAP and CAS) of produce is mostly credited to ozone-instigated structural changes in their cellulose and hemicellulose constituents. Celluloses and hemicelluloses are the two chief components of plant walls that endure ozone-prompted polymerization and epimerization that encourage congealing of the cell walls. Consequently, this led to notable changes in the texture of freshly cut vegetables, such as asparagus, coriander, and spinach.

Per Skog and Chu (2001), the crunchiness of cucumbers and firmness of citrus fruits was found to be significantly enhanced by ozone processing but at minimal ozone concentrations ranging from 0.1 to 10 mg/L and short treatment times, ranging from three to five minutes. Prolonged exposure to ozonated water results in considerable oxidation of feruloylated or phenolic cross-linkages in cell-wall pectin, structural proteins, or other polymers that in turn change the texture of the product to soft and soggy.

4.10.4 Sensorial Characteristics of the Food System

Owing to the fact that the majority of aromatic compounds are unsaturated in chemical nature, such as ethyl acetate, ethyl butyrate, ethyl butanoate, isoamyl acetate, pentyl butyrate, and pentyl butanoate, one of the most prominent impacts is molecular ozone-induced damage of aromatic compounds. This structural damage to organoleptic compounds significantly alters the sensorial characteristics of the ozone-processed food commodities. For instance, in the case of ozone-supplemented cold storage, strawberries suffered an irreversible loss of fruit aroma over the course of the storage period. The major reason behind this loss of aroma was found to be the considerable oxidation of aromatic compounds, in particular those contributing to the fruity flavor of the product. Many researchers have noted that in order to accomplish effective decontamination of produce, the utilization of high ozone dosages and high exposure time explicitly alters the organoleptic characteristics of commodities.

4.11 Conclusion

The demand for nutritious and high-quality food items is increasing as the modern world's health consciousness grows. Sanitation/disinfection is a critical function to ensure the microbiological safety of processed and stored foods. Because of the formation of potentially carcinogenic by-products in food applications, conventionally used sanitizing agents based on chlorine and bromine are under examination, prompting severe laws across the world, particularly in industrialized nations. As a result, the food business is on the lookout for a powerful replacement. Ozone is a powerful oxidizing agent that has the ability to inactivate microorganisms and mycotoxins on food surfaces. It has been found that it is as effective as, if not more effective than, chlorine levels during aqueous sanitization of fresh fruits and vegetables. Because of its very unstable nature, the gas must be created on site, eliminating the requirement for transportation and the risks during storage. It leaves no residue and is extremely effective in removing pesticides and other pollutants from food. Ozone may be employed in both gaseous and aqueous forms in a wide range of food products, including fresh fruits and vegetables, juices, processed meals, cereals, and so on.

References

Abad, E. X.; Yinto, R. M.; Gajardo, R.; Bosch, A. Viruses in mussels: Public health implications and depuration. *J. Food Prot.* 1997, 60, 677–681.

Akbas, M. Y.; Ozdemir, M. Effectiveness of ozone for inactivation of *Escherichia coli* and *Bacillus cereus* in pistachios. *Int. J. Food Sci. Technol.* 2006, 41, 513–519.

Alexopoulos, A., Plessas, S., Ceciu, S., Lazar, V., Mantzourani, I., Voidarou, C., . . . & Bezirtzoglou, E. (2013). Evaluation of ozone efficacy on the reduction of microbial population of fresh cut lettuce (Lactuca sativa) and green bell pepper (Capsicum annuum). *Food control*, 30(2), 491–496.

An, J., Zhang, M., & Lu, Q. (2007). Changes in some quality indexes in fresh-cut green asparagus pretreated with aqueous ozone and subsequent modified atmosphere packaging. *Journal of Food Engineering*, 78(1), 340–344.

Bablon, G.; Bellamy, W. D.; Bourbigot, M-M.; Daniel, E. B.; Dore, M.; Erb, E.; Gordon, G.; Langlais, B.; Laplanche, A.; Legube, B.; Martin, G.; Masschelein, W. J.; Pacey, G.; Reckhow, D. A.; Ventresque, C. Fundamental aspects. *In Ozone in Water Treatment, Application and Engineering*, pp. 11–132. Lewis Publishers, Inc., Chelsea, MI, 1991.

Beltrán, F. J. (2003). Ozone Reaction Kinetics for Water and Wastewater Systems. Boca Raton, FL: Lewis Publishers/CRC Press.

Bezbarua, B. K., & Reckhow, D. A. (2004). Modification of the standard neutral ozone decomposition model. *Ozone: Science & Engineering*, 26(4), 345–357.

Chen, C.; Liu, C.; Jiang, A.; Zhao, Q.; Liu, S.; Hu, W. Effects of ozonated water on microbial growth, quality retention and pesticide residue removal of fresh-cut onions. *Ozone Sci. Eng.* 2020, 42, 399–407.

Chen, H. C.; Huang, S. H.; Moody, M. V.; Jiang, S. T. Bacteriocidal and mutagenic effects of ozone on shrimp *(Penaeus monodon)* meat. *J. Food Sci.* 1992, 57, 923–927.

Cox, J. E.; Cox, J. M.; Duffy Cox, R. W. Hyperpasteurization of food. US Patent 5 431 939, 1995.

Dave, S. Efficacy of ozone against *Salmonella enteritidis* in aqueous suspensions and on poultry meat. M.Sc. Thesis, The Ohio State University, Columbus, OH, 1999.

DeWitt, B. J.; McCoid, V.; Holt, B. L.; Ellis, D. K.; Finne, G.; Nickelson, R. Proceedings of the ninth annual tropical and subtropical fisheries technological conference of the Americas, 9–12 January, pp. 260–279. The potential use of ozonated ice for on-board storage of Gulf of Mexico shrimp. Brownsville, TX, 1984.

Diaz, M. E.; Law, S. E. Proceedings of the 14th Ozone World Congress, IOA, UV-enhanced ozonation for reduction of pathogenic microorganisms and turbidity in poultry-processing chiller water for recycling, pp. 391–403. Dearborne, MI, 1999.

Dolly, S. B.; Singh, A.; Sharma, S. Ozone as a shelf-life extender of fruits. *In Emerging Technologies for Shelf-Life Enhancement of Fruits*, pp. 289–312. Apple Academic Press, Florida, USA. 2020.

Fábián, I. (2006). Reactive intermediates in aqueous ozone decomposition: A mechanistic approach. *Pure and Applied Chemistry*, 78(8), 1559–1570.

Ferreira, W. F. D. S., Alencar, E. R. D., Alves, H., Ribeiro, J. L., & Silva, C. R. D. (2017). Influence of pH on the efficacy of ozonated water to control microorganisms and its effect on the quality of stored strawberries (Fragaria x ananassa Duch.). *Ciência e Agrotecnologia*, 41, 692–700.

Fukuzaki, S. The use of gaseous ozone as a cleaning agent on stainless steel surfaces fouled with bovine protein. *Ozone: Sci. Eng.* 2006, 28, 303–308.

Gabler, F. M., Smilanick, J. L., Mansour, M. F., & Karaca, H. (2010). Influence of fumigation with high concentrations of ozone gas on postharvest gray mold and fungicide residues on table grapes. *Postharvest Biology and Technology*, 55(2), 85–90.

Gardoni, D., Vailati, A., & Canziani, R. (2012). Decay of ozone in water: A review. *Ozone: Science & Engineering*, 34(4), 233–242.

Galdeano, M. C., Wilhelm, A. E., Goulart, I. B., Tonon, R. V., Freitas-Silva, O., Germani, R., & Chávez, D. W. H. (2018). Effect of water temperature and pH on the concentration and time of ozone saturation. *Brazilian Journal of Food Technology*, 21.

Gottschalk, C., Libra, J. A., & Saupe, A. (2009). *Ozonation of Water and WasteWater. A Practical Guide to Understanding Ozone and its Application. 2nd Edition.* New York: Wiley-VCH.

Guzel-Seydim, Z. B.; Greene, A. K.; Seydim, A. C. Use of ozone in the food industry. *LWT—Food Sci. Technol.* 2004, 37, 453–460.

Guzel-Seydim, Z. B.; Wyffels, J. T.; Greene, A. K.; Bodine, A. B. Removal of dairy soil from heated stainless steel surfaces: Use of ozonated water as a prerinse. *J. Dairy Sci.* 2000, 83, 1887–1891.

Haraguchi, T.; Simidu, U.; Aiso, K. Preserving effect of ozone to fish. *BulL J. Soc. Sci. Fish.* 1969, 35, 915–919.

Horvath, M.; Bilitzky, L.; Huttner, J. *Ozone.* Elsevier, New York, 1985.

Ignatiev, A. N., Pryakhin, A. N., & Lunin, V. V. (2008). Numerical simulation of the kinetics of ozone decomposition in an aqueous solution. *Russian Chemical Bulletin*, 57, 1172–1178.

Ingram, M.; Haines, R. B. Inhibition of bacterial growth by pure ozone in the presence of nutrients. *J. Hyg. (Cambr.).* 1949, 47, 146–158.

Izat, A. L.; Adams, M.; Colberg, M.; Reiber, M. Effects of ozonated chill water on microbiological quality and clarity of broiler processing water. *Arkansas Farm Res.* 1990, 39(2), 9–18.

Jindal, V.; Waldroup, A. L.; Forsythe, R. H.; Miller, M. Ozone and improvement of quality and shelflife of poultry products. *J. Appl. Poul. Res.* 1995, 4, 239–248.

Jung, Y., Hong, E., Kwon, M., & Kang, J. W. (2017). A kinetic study of ozone decay and bromine formation in saltwater ozonation: Effect of O_3 dose, salinity, pH, and temperature. *Chemical Engineering Journal*, 312, 30–38.

Jurado-Alameda, E.; Altmajer-Vaz, D.; Garca-Roman, M.; Jimenez-Perez, J. L. Study of heat-denatured whey protein removal from stainless steel surfaces in clean-in-place systems. *Int. Dairy J.* 2014, 38, 195–198.

Karaca, H.; Velioglu, Y. S. Effects of ozone and chlorine washes and subsequent cold storage on microbiological quality and shelf life of fresh parsley leaves. *LWT.* 2020, 127, 109421. doi: 10.1016/j. lwt.2020.109421.

Kells, S. A.; Mason, L. J.; Maier, D. E.; Woloshuk, C. P. Efficacy and fumigation characteristics of ozone in stored maize. *J. Stored Prod. Res.* 2001, 37, 371–382.

Khadhraoui, M., Trabelsi, H., Ksibi, M., Bouguerra, S., & Elleuch, B. (2009). Discoloration and detoxification of a Congo red dye solution by means of ozone treatment for a possible water reuse. *Journal of Hazardous Materials*, 161(2–3), 974–981.

Khadre, M. A.; Yousef, A. E. Decontamination of a multi-laminated aseptic food packaging material and stainless steel by ozone. *J. Food Safety.* 2001, 21, 1–13.

Kim, J. G. Ozone as an antimicrobial agent in minimally processed foods. Ph.D. Dissertation, The Ohio State University, Columbus, OH, 1998.

Kim, J. G., Yousef, A. E., & Khadre, M. A. (2003). Ozone and its current and future application in the food industry. *Adv Food Nutr Res.*, 45: 167–218.

Lee, J. S.; Kramer, D. E. Effectiveness of ozone-treated wash water and ice on keeping quality and stability of Sockeye salmon. Report FITC 84/T-1 to Alaska Sea Grant College Program and Alaska Department of Environmental Conservation, July 1984.

Liu, C.; Chen, C.; Jiang, A.; Zhang, Y.; Zhao, Q.; Hu, W. Effects of aqueous ozone treatment on microbial growth, quality, and pesticide residue of fresh-cut cabbage. *Food Sci. Nutr.* 2021, 9, 52–61.

Liu, C., Ma, T., Hu, W., Tian, M., & Sun, L. (2016). Effects of aqueous ozone treatments on microbial load reduction and shelf life extension of fresh-cut apple. *International Journal of Food Science & Technology*, 51(5), 1099–1109.

McKenzie, K. S.; Sarr, A. B.; Mayura, K.; Bailey, R. H.; Miller, D. R.; Rogers, T. D. Oxidative degradation and detoxification of mycotoxins using a novel source of ozone. *Food Chem. Toxicol.* 1997, 35, 807–820.

Mitsuda, H.; Ominami, H.; Yamamoto, A. Synergistic effects of ozone and carbon dioxide gases for sterilizing food. *Proc. Japan. Acad.* 1990, 66, 68–72.

Mimura, G., Katayama, Y., Ji, X., Xie, J., & Namba, K. (1998). Acute toxicity of ozone-exposed seawater and chlorinated seawater for Japanese flounder, Paralichthys olivaceus, eggs, larvae and juveniles. *Aquaculture Science*, 46(4), 569–578.

Mosayebi, B. D.; Zokaie, N. Extension of fish shelf life by ozone treatment. *Int. J. Environ. Ecol. Eng.* 2010, 4, 220–230.

Nakano, Y., Okawa, K., Nishijima, W., & Okada, M. (2003). Ozone decomposition of hazardous chemical substance in organic solvents. *Water Research*, 37(11), 2595–2598.

Pandiselvam, R.; Singh, A.; Agriopoulou, S.; Sachadyn-Król, M.; Aslam, R.; Lima, C. M. G.; Khanashyam, A. C.; Kothakota, A.; Atakan, O.; Kumar, M.; Mathanghi, S. K. A comprehensive review of impacts of ozone treatment on textural properties in different food products. *Trends Food Sci Technol.* 2022, 127, 74–86.

Patil, S.; Valdramidis, V. P.; Cullen, P. J.; Frias, J.; Bourke, P.; Inactivation of Escherichia coli by ozone treatment of apple juice at different pH levels. *Food Microbiol.* 2010, 27, 835–840.

Psaltou, S.; Mitrakas, M.; Zouboulis, A. Catalytic Membrane Ozonation. *Encyclopedia*, 2021, 1, 131–143.

Rice, R. G., Application of ozone in water and waste water treatment. *In* Rice, R. G.; Browning, M. J. (Eds.), *Analytical Aspects of Ozone Treatment of Water and Waste Water*, p. 726. Syracuse, The Institute, New York, 1986.

Rojek, U.; Hill, A. R.; Griffiths, M. Preservation of milk by hyperbaric ozone processing. *J. Dairy Sci.* 1995, 78, 115–125.

Sander, M. Process for the gentle ozone treatment of liquids, such as fruit juices, milk, liquid milk products, wine, oils, liquid medicaments, blood and/or similar products (In German). Patent No. DE 3325568A1, 1985.

Scott, D. B. M.; Lesher, E. C. Effect of ozone on survival and permeability of *Escherichia coli. J. Bacteriol.* 2004, 85, 567–576.

Segat, A.; Misra, N. N.; Fabbro, A.; Buchini, F.; Lippe, G.; Cullen, P. J.; Innocente, N. Effects of ozone processing on chemical, structural and functional properties of whey protein isolate. *Food Res. Int.* 2014, 66, 365–372.

Sharma, R.; Bhandari, M.; Kaur, K.; Singh, A.; Sharma, S.; Kaur, P. Molecular interactome & starch-protein matrix, functional properties, phytochemical constituents and antioxidant activity of foxtail millet (*Setaria italica*) flour as influenced during gaseous ozonation. *Cereal Chem.* 2022, 99(5), 1101–1111.

Sharma, R.; Singh, A.; Sharma, S. Influence of ozonation on cereal flour functionality and dough characteristics: A review. *Ozone Sci. Eng.* 2021, 43(6), 613–636.

Staehelin, J., & Hoigne, J. (1982). Decomposition of ozone in water: rate of initiation by hydroxide ions and hydrogen peroxide. *Environmental Science & Technology*, 16(10), 676–681.

Staehelin, J; Hoignr, J. Ozone decomposition in water studied by pulse radiolysis. 2. OH and HO 4 as chain intermediates. *J. Phys. Chem.* 1985, 88, 5999–6004.

Tomiyasu, H.; Fukutomi, H.; Gordon, G. Kinetics and mechanisms of ozone decomposition in basic aqueous solution. *Lnorg. Chem.* 1985, 24, 2962–2966.

Von Gunten, U. (2003). Ozonation of drinking water: Part I. Oxidation kinetics and product formation. *Water Research*, 37(7), 1443–1467.

Waldroup, A. L.; Hierholzer, R. E.; Forsythe, R. H.; Miller, M. J. Recycling of poultry chill water using ozone. *J. Appl. Poul. Res.* 1993, 2, 330–336.

Wang, J.; Wang, S.; Sun, Y.; Li, C.; Li, Y.; Zhang, Q.; Wu, Z. Reduction of Escherichia Coli O157:H7 and naturally present microbes on fresh-cut lettuce using lactic acid and aqueous ozone. *RSC Adv.* 2019, 9, 22636–22643.

Whistler, P. E.; Sheldon, B. W. Biocidal activity of ozone versus formaldehyde against poultry pathogens inoculated in a prototype setter. *Poult. Sci.* 1989, 68, 1068–1073.

Wu, J., Doan, H., & Cuenca, M. A. (2006). Investigation of gaseous ozone as an anti-fungal fumigant for stored wheat. *Journal of Chemical Technology & Biotechnology: International Research in Process, Environmental & Clean Technology*, 81(7), 1288–1293.

Wysok, B., Uradziñski, J., & Gomólka-Pawlicka, M. (2006). Ozone as an alternative disinfectant – A review. *Polish Journal of Food and Nutrition Sciences*, 15(1), 3.

Yousef, A. E.; Rodriguez-Romo, L. Methods for decontaminating shell eggs. U.S. Patent Application, Dockett Number: 22727-04099, 2001.

Chapter 5

Bio-Based Composites for Food Packaging

L. Muthulakshmi, S. Mohan, and R. Rajam

5.1 Introduction

Food packaging could be regarded as a casing that helps in the proper protection of food materials from various factors that are physical, chemical and microbial, deeply influencing the type of packaging procedure to be followed (Alamri et al., 2021). Moreover, food packaging has proven to be a monitor for maintaining the quality of food, provide traceability to analyse its flow, extend the shelf life and prevent any adulterations (Marsh & Bugusu, 2007). The utilization of food packaging materials began as early as the Industrial Revolution and world wars to supply food for those living in distant locations and many other adversary conditions (Risch, 2009). The most commonly used packaging materials for food products include paper, metal, glass and cardboard. Despite some of them having sustainable sources, processing them into usable forms requires a huge expenditure of energy. So bio-based products were highlighted due to their biodegradable nature (Berk, 2013; Halonen et al., 2020). Bio-based polymers are employed for sustainable or green packaging (SOGP) material that helps establish a proper sustainable source for the fabrication of food packaging materials (Wang et al., 2021). However, to meet sustainable goals, a deeply interwoven network between politics, academia and research is required for proper implementation of the sustainable material readily available for the consumer (Reichert et al., 2020). However, bio-based materials, when

DOI: 10.1201/9781003217138-5

deployed individually for the development of polymers, do not have the desirable properties required for ideal packaging material, so biocomposites could be used as a replacement for them. Biocomposites are polymer materials that are formed as a combination between one or more individual polymers, mainly to improve the characteristic features of the biocomposite material (Rudin & Choi, 2012). Biocomposites are produced from many sources, like carbohydrates, proteins and lipids. Polysaccharides are one of the most viable sources of biocomposites due to their abundance in the atmosphere. Several sources of bio-based materials offer a huge range of polysaccharides with the desired properties to be used for the manufacture of packaging material (Wahab & Abd Razak, 2016). Carrageenan is one of many polysaccharides that are sulphated and offer good, flexible properties to be used as a bio-polymer candidate, and it has the ability to sulphate other copolymers that could be used for the synthesis of packaging material (Bhat et al., 2020). Alginate is one of the most widely used and studied polysaccharides and has been analysed for its ability to be fabricated as a packaging material (Senturk Parreidt et al., 2018). Similarly, polysaccharides like starch, cellulose, pullulan and xanthan gum tend to produce efficient materials that are capable of protecting foodstuffs. Proteins also play a major part in the development of biocomposites. They have been found more suitable for their role as edible packaging material. Proteins like zein, gluten, gelatin and soy protein and milk-soluble proteins like whey and casein have been explored for their potential to form biocomposites with improved properties to be used as developing food packaging material (Cheng et al., 2019). Lipids like fat and oil are also equally studied due to their inherent gas barrier properties (Yousuf et al., 2021). Smart packaging using biocomposites is a growing field that requires attention, as both consumer-friendly and sustainable solutions could be achieved simultaneously to produce better food packaging material.

Thus, this chapter outlines the role of food packaging; the forms and materials used for food packaging; role of bio-based compounds for the manufacture of food packaging; biocomposites of polysaccharides, proteins and lipids used for synthesis of food packaging; and biocomposites used in active food packaging material.

5.2 Food Packaging

Food packaging is used to enclose a wide variety of food products to ensure protection of the food material. The cost of handling would be increasingly high without proper packaging, and unpacked foods would spoil within a short span of time. The preservation of food material is ensured against a variety of factors like microbial contamination and physical and chemical damage. The major considerations in deciding on a suitable packaging material for food include the quality, cost and shelf life (Alamri et al., 2021). The need for packaging beyond food protection and preservation involves food wastage reduction by containment, a vector of information for marketing, traceability for ensuring the food supply chain, handling convenience and preventing adulterations, and packaging could be used as household containers (Marsh & Bugusu, 2007). Proposed food packaging materials are tested robustly by food safety organizations like the European Commission, Food and Drug Administration and Brazil National Health Surveillance Agency. The wide use of food packaging materials began as early as the 18th century, and they have greatly grown in use since then (Han et al., 2018). The early Industrial Revolution brought development to many products and their delivery schemes. Although it was not exclusive for food products, over time, the focus on packed delivery of food materials became widespread. The rate of food packaging increased during the world wars, helping consumers to eat food materials at the time (Risch, 2009). Packaging has a direct appeal to the consumer and can showcased the product. In other words, packaging overlaps with dynamic, complex, scientific and other controversial segments of business. So packaging could be portrayed as having a techno-economic function that aims at reducing the cost of the product while proportionally increasing sales and profits. The facts influencing the design of packaging material are the nature of the product, costs associated with distribution, requirements with respect to the market and selection of material and machinery (Paine & Paine, 2012).

5.4 Materials Used for Food Packaging

The most common types of food packaging can be flexible or rigid in nature, with a number of levels to promote ease in transportation. Forms of food packaging might include cans, wrappers, sleeves, jars and cartons (Berk, 2018). The materials used for packaging are described in the following, and some are shown in Figure 5.1.

   5.4.1 Metals: Metals possess increased mechanical strength, fine thermal conductivity and heat resistance. Steel and aluminium re widely deployed for food packaging, as 9% of packaging material.

   5.4.2 Glass: Glass offers rigidity, thermal resistance, inertness and transparence. Recycling and reuse of glass containers is also possible. Around 11% of food packaging material is glass.

   5.4.3 Paper: It is the most primitive type of packaging material, with 34% of the total food packaging material. It can be produced with low cost and is printable for including information about the product.

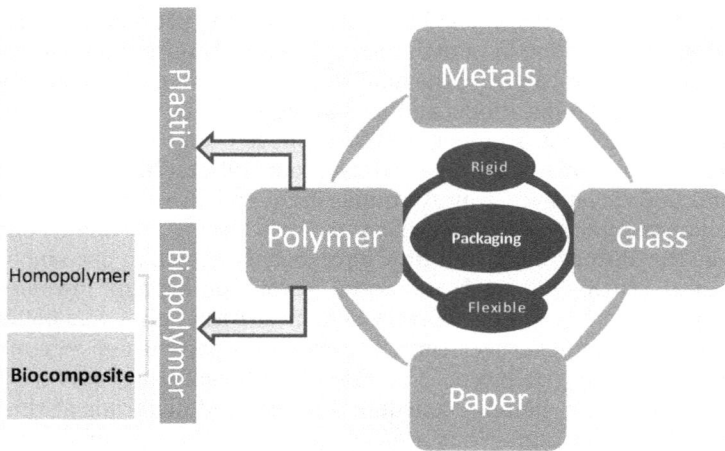

Figure 5.1 Materials used for food packaging.

5.4.4 Polymer: Polymers have evolved to be a better option for developing packaging material due to their flexible nature and have occupied a major portion of the current food packaging materials, representing 37% of the whole (Berk, 2013; Halonen et al., 2020).

5.4 Bio-Based Materials for Food Packaging

Plastic pollution is one of the alarming environmental concerns that have to be addressed to prevent further damage. An exponential increase in the amount of plastic waste is estimated by 2050, most of which will be contributed by plastic used for food packaging (Porta et al., 2022). Commonly used plastics in food packages include polypropylene, low-density polyethylene, high-density polyethylene and polyethylene terephthalate, accounting for nearly 39.9% of the entire share of plastic produced. As a replacement for conventional plastic polymers, bio-based polymers could be used, as they are biodegradable in nature (Halonen et al., 2020). Biopolymers could be extracted from many sources, like bacteria, fungi, biomass and other natural material. Generally based on their origin, biopolymers are classified as those from carbohydrate, proteins or a copolymer of both (Asgher et al., 2020). Biopolymers are an important part of sustainable or green packaging, which focuses on the source

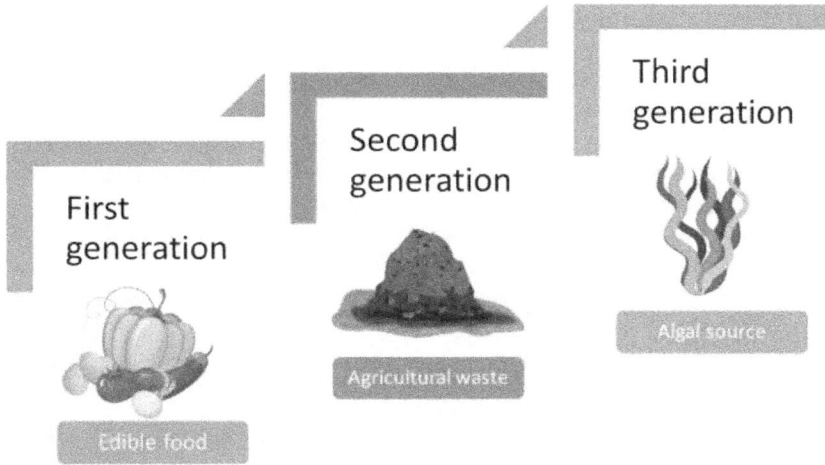

Figure 5.2 Generations of packaging material based on the sources of raw materials.

material, the process adopted for production and the management of waste. In short, SOGP emphasizes raw materials from renewable sources, economical feasibility and biodegradability of the material. So bio-based products fit into these aspects of SOGP more aptly than conventional plastics (Wang et al., 2021). Based on the reservoir of raw material used for the fabrication of packaging material, generation of feedstock could be categorized as first generation, in which the biomass is edible food; second generation, which includes materials from agriculture, forests and municipal waste; and third generation, which uses algae for the production of biopolymers, as shown in Figure 5.2.

Promoting the ideal of sustainability that would eventually have a domino effect on politics, industry and academia could be accomplished by biopolymers in an effective way (Reichert et al., 2020).

5.5 Biocomposites

Biocomposites are materials that are made up of a combination of two or materials, at least one of which is of natural origin, to produce a novel substance that has the capability of improved performance over the individual constituents (Rudin & Choi, 2012). Methods for increasing the applicability of polymers include using fillers, blending with other polymers or using plasticizers (Zafar et al., 2016).

5.5.1 Polysaccharide-Based Biocomposites

Polysaccharides are complexes of monosaccharide units bound together by glycosidic linkages. In addition to being abundant, they are biocompatible and non-toxic in nature. The unreliable physical properties of polysaccharides could be enhanced by modifying their properties by fabricating them as composite materials (Wahab & Abd Razak, 2016). Some of the polysaccharides developed as biocomposites are discussed in this section and shown in Figure 5.3.

5.5.2 Carrageenan Biocomposites

Carrageenan is a unique water-soluble polymer of sugar units encompassed with sulphate groups. It is most commonly obtained from the cell wall of red seaweed. The properties of extracted carrageenan depend mainly on the amount of sulphate clusters, physio-chemical properties and its structure (Bhat et al., 2020). Carrageenan films are usually created by a

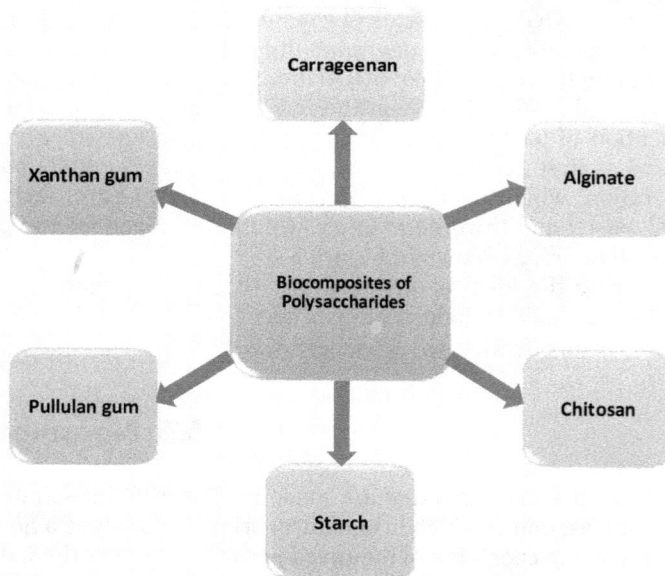

Figure 5.3 Biocomposites of polysaccharides.

simple procedure in which neutral or alkaline solution at a hot temperature is allowed to gelate, followed by drying. Due to their hydrophilic nature, despite being a poor barrier for moisture, they are better in halting the entry of oxygen and lipid (Sanchez-García, 2011). A biocomposite of carrageenan with pectin reduced water permeability further on complexing with nanoclay as filler (Alves et al., 2011). Carrageenan in a complex with silica and zinc particles blended with cassava starch yielded a stable polymer film that is edible (Praseptiangga et al., 2021). Carrageenan polymers are used as edible films over the desserts, chicken or any other product requiring an edible coating (Vishnuvarthanan & Rajeswari, 2019).

5.5.3 Alginate Biocomposites

Alginate is one of the naturally occurring polysaccharides that can be extracted from brown algae. Alginate is extracted in the form of alginic acid as complexes with sodium, potassium, calcium and ammonium. It can be moulded into various forms per requirements (Senturk Parreidt et al., 2018). It is mostly used as an agent for thickening, stabilization, film forming and gel-producing. Alginate can be combined with synthetic substances like polyvinyl alcohol; carbon nanofibers; or natural materials like cellulose, oil, cotton and chitosan (Abdullah et al., 2021). Alginate is suitable for the design of edible packaging. On the addition of cellulose, glycerol, calcium and sodium alginate, waterproof packaging material is prepared that has a hydrophobic surface. The morphological structure is glossy and flexible, with a smooth surface that is devoid of cracks (Žiūkaitė et al., 2022). Generally, alginate is hydrophilic in nature, which leads to its poor water barrier properties. In order to improve the water barrier property, essential oil is used as a plasticizer along with alginate to form edible films. These coatings also exhibit antibacterial properties due to the incorporation of edible oil (Frank et al., 2018).

5.5.4 Chitosan Biocomposites

Chitosan is derived from chitin, which is the second-largest group of polysaccharides available on the planet. It is usually obtained from multiple sources, like seafood, making it cheap and one of the commercially available polysaccharides. It is crystalline and is soluble only in acidic solvents (Flórez et al., 2022). Chitosan possesses inherent antimicrobial properties, while improving its mechanical and barrier properties could be done by employing other polymers in combination with it. Chitosan coatings have been found to retard the ripening of fruits and vegetables, synergistically expanding their stability periods (Souza et al., 2020). Chitosan can be made into films in its pure form or a complex with other polymers, polysaccharides, proteins, extracts, synthetic polymers, inorganic materials and derived products. It is produced by methods like direct casting, coating, spread coating, spray coating, immersing or dipping, assembly by layer and extrusion (Wang et al., 2018). It is found that the browning index of pineapples can be considerably reduced using chitosan complexed with starch along with an additional plasticizer of pineapple leaf microfibres with increased ability to degrade by itself (Mutmainna et al., 2019).

5.5.5 Starch Biocomposites

Starch is one of the most common sources of polysaccharides and is obtained from potato, wheat, rice and corn. It is a semi-crystalline molecule buildup of units like amylose and amylopectin. The proportion or ratio of amylose and amylopectin depends on the source of material used for the extraction of starch (Mallick et al., 2020). It is not thermoplastic in nature, but this property can be improved by the use of a plasticizer. Upon heating, disruption of the structure and diffusion of the plasticizer occurs, leading to its sensitivity to moisture. The modification of native starch can be done by thermos-plasticization; chemical modifications like heat moisture, super-heated steam, high hydrostatic pressure or microwave heat; and physical modifications using ultrasound, UV irradiation, corona electric discharge and pulsed electric fields (García-Guzmán et al., 2022). The main sources of starch include corn, potato, wheat, cassava and tapioca. The most common disadvantages of starch are poor mechanical strength and increased permeability to water vapor (Onyeaka et al., 2022). In order to solve this issue, starch is often converted to biocomposites, with other polymeric materials being plasticized using suitable plasticizers (Mustapha et al., 2019).

5.5.6 Pullulan Biocomposites

Pullulan is an exopolysaccharide produced by microbes that are water soluble and non-toxic with film-forming and adhesive properties, and it is capable of forming transparent films with considerable mechano-physical properties (Kraśniewska et al., 2019). Pullulan copolymerizing with an apple fibre–fabricated biocomposite films possessed better antimicrobial, antioxidant properties and significantly increased the property of contact angle (Luis et al., 2021). Pullulan and chitosan-based material is reinforced with zinc nanoparticles to increase the antimicrobial activity of the films without influencing the transparency of the film used for coating pork belly to prevent microbial contamination (Roy et al., 2021).

5.5.7 Xanthan Gum Biocomposites

This is a microbial exopolysaccharide produced by *Xanthomonas* species with increased solubility of water and biocompatibility (Kumar et al., 2018). The biocomposite of xanthan gum and curdlan gum offered improved properties of the material that could be used as a film for coating food material. The films produced from the blend polymer have significant barrier and mechanical properties (Mohsin et al., 2020). On the incorporation of titanium dioxide and silver nanoparticles along with lemon peel powder, xanthan gum films showed antibiotic activity against prominent foodborne pathogens (Meydanju et al., 2022).

5.6 Protein-Based Biocomposites

Compared to other macromolecules, protein-based biopolymers are more useful due to their better gas barrier properties. They are considered a critical candidate in the synthesis of edible films that can be used as a packaging material for food due to their intrinsic properties (Cheng et al., 2019). Some of the proteins used for manufacture of food packaging material are described in this section and shown in Figure 5.4.

5.6.1 Gluten Biopolymers

Gluten is one of the most abundantly available protein in wheat sources. It has been found that the bioplastics produced from wheat gluten are nontoxic and completely biodegradable in nature. The properties of the gluten matrix can be enhanced by synthesizing gluten as a biocomposite with other polymers (Patni et al., 2014). Wheat gluten–based biocomposites are most often synthesized at the nanoscale with paper coating for atmospheric packaging for fruits and vegetables with improvised mass transfer and mechanical properties (Angellier-Coussy et al., 2011). When thermoplastic gluten is polymerized with polycaprolactone (PCL) in addition to chrome octanoate, the resulting film is found to possess shape memory and is completely compostable in nature (Gutiérrez et al., 2021).

5.6.2 Soy Protein Biopolymers

Soy protein is one of the most common proteins derived from plant sources with the ability to be utilized as a sustainable source for food packaging. It is known to be biocompatible and biodegradable in nature. It also has a lower

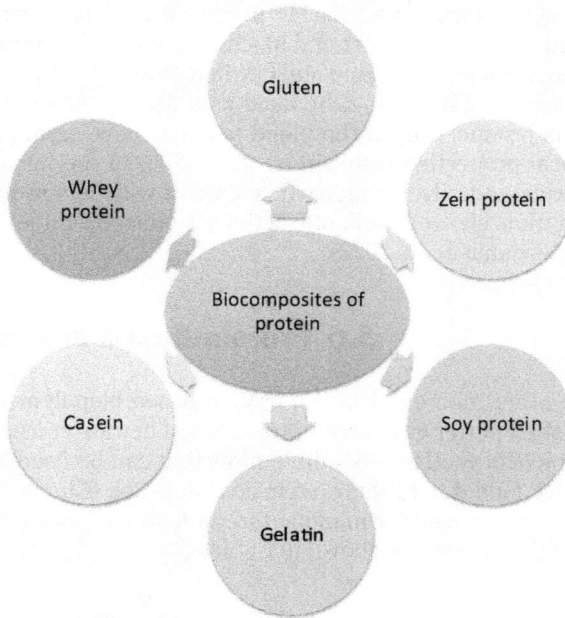

Figure 5.4 Biocomposites of protein.

cost for production due to its low water resistance and strength (Huang et al., 2021). Soy protein in combination with polylactic acid produces bilayers that are transparent and strongly adhesive with each other. The addition of PLA improved the properties associated with water vapour and soluble matter. On incorporation with an antibiotic agent, a bilayer film inhibited the growth of many food contaminants (González & Igarzabal, 2013). At the nanoscale, soy protein biocomposites have improved properties of gas barriers, processibility and film-forming, emphasizing their potential as a suitable packaging material (Prusty et al., 2021). When soy protein films are incorporated with cellulose nanofibril and lactic acid, along with other plant extracts, then the film's antimicrobial ability is evident, with action against many foodborne pathogens. They also have antioxidants and light barrier ability (Yu et al., 2019).

5.6.3 Zein Protein Biopolymers

Zein is a soluble protein extracted from corn with a hydrophobic nature due to the increased presence of non-polar amino acids, and it is known to have thermoplastic properties that make it capable of forming films (Bayer, 2021). It is obtained as a by-product in the processing of corn and maize in the oil and bioethanol industries. It also has many natural active agents that are antibiotic and antioxidant in nature (Arcan et al., 2017). Zein protein is categorized as generally regarded as safe (GRAS) for the encapsulation of bioactive compounds by the Food and Drug Administration (FDA). It is claimed to be a sustainable source for the production of nanomaterials that could effectively meet the demands of the consumer (Malhotra & Alghuthaymi, 2022). Zein and chitosan are some of the biopolymers from renewable sources that might have improved properties of elongation with varying strengths of the film. Due to their fragile nature, they are copolymerized with chitosan to allow flexibility, transparency and potent oxygen barrier properties (Bueno et al., 2021). When zein protein is mixed with xanthan gum, water solubility and opacity increase, with a simultaneous increase in tensile strength (Pena Serna & Lopes Filho, 2015).

5.6.4 Gelatin Biopolymers

Gelatin is a partial digest of collagen from animal tissue that has wide applications in the food, photography, cosmetic and pharmaceutical industries. It has good water-binding, gel-forming, film-forming and foam-forming properties, along with a permissible gas barrier (Lu et al., 2022). The soluble form of gelatin is usually obtained by the destabilization of the triple helix of collagen, which is influenced by factors like the initial characteristics of the collagen and the process followed for extracting it. Usually destabilization occurs by breaking down the bonds like hydrogen and covalent bonds, mostly by heat treatment (Ramos et al., 2016). Gelatin sulphated with carrageenan incorporated with nanoemulsions of polyphenols with antimicrobial activity is found to extend the shelf life of broiler meat to a considerably longer period than the conventional procedures followed for packaging (Khan et al., 2020). The protein hydrolysate from shrimp and crab added to composite films of chitosan and gelatin showed better UV-barrier, hydrophilicity, antioxidant and antibacterial properties (Hajji et al., 2021). Protein isolates from chicken combined with gelatin of fish skin can be incorporated with phenolic compounds of importance to produce active packaging material for storing chicken skin oil (Nilsuwan et al., 2021).

5.6.5 Casein Protein Biopolymers

Around 80% of the total protein in milk is casein present as by-products in many dairy products in complex with calcium and sodium ions. Due to the presence of an increased number of polar functional groups, casein protein possesses a better barrier to oxygen, making it suitable for edible food packaging (Bonnaillie et al., 2014). This protein is biodegradable and has the ability to form films, as it has random coils and the ability to form hydrophobic, electrostatic and hydrogen bonds randomly, and it is known to have antimicrobial effects (Saez-Orviz et al., 2017). In addition to natural taste and flavour, the milk proteins are highly suitable to design packaging materials for specific food categories due to their specific characteristics such as moisture loss prevention, barrier to oxygen, good tensile strength, flexibility and elongation. Also, the structure and the functionality of milk proteins are suitable to produce edible polymer which can be used to incorporate, transport and deliver the nutraceuticals for food and biotechnological applications (Daniloski et al., 2021). Casein films are generally fabricated by casting and extrusion, while coating forms are produced by solution casting. Casein is known to be complexed with Arabic gum, chitosan, potato starch and many other polymers to improve the properties of the casein (Khan et al., 2021).

5.6.6 Whey Protein Biopolymers

Whey proteins offer good solubility and have better emulsifying ability, with complete biodegradability and an appetizing nature. Whey-protein based coating has been used as a coating for many food materials like nuts, cereals, fruits and frozen food. It makes up to 85–95% of the volume of milk, with about 55% of the nutrients (Kandasamy et al., 2021). Whey proteins are known to offer better aroma, humidity, fat and oxygen barriers. Therefore, whey protein coatings could exclusively retard the oxidation of lipids, extending the shelf life of food material by preventing rancidity (Bugnicourt et al., 2013). When montmorillonite nanoparticles are incorporated into whey protein isolate, the mechanical and barrier properties are improved. The use of plasticizers like glycerol and sorbitol greatly decreased the permeability of oxygen within the coating material (Schmid et al., 2017). Zein protein nanoparticles on a coating with sodium caseinate increased the mechanical and barrier properties of the film without influencing the elongation property and decreasing the hydrophilic nature of the material (Oymaci & Altinkaya, 2016).

5.7 Lipid-Based Biocomposites

Lipids from plants and animals include fats and oils, respectively. Moreover, their structures are similar, but the physical forms of both vary, with the former being solid and the latter being liquid in nature (Mohamed et al., 2020). The most common fatty acids with carbon numbers between 14 and 18 in the form of waxes, fatty alcohol and vegetable oil are used for the synthesis of packaging material (Aydin et al., 2017). Lipids are also a vital part of edible coating materials and are biodegradable in nature (Yousuf et al., 2021).

5.8 Active Packaging of Food with Biocomposites

In order to enhance the quality of active packaging, many fillers and subsidiary constituents are deliberately added by adopting different standard procedures (Lim, 2015). It is considered a suitable solution for many applications in the food industry. At present, the active packaging materials, mostly used in food industry to increase the shelf life of food, oxygen and other molecular scavenging activity, regulate ethylene and provide antimicrobial agents to prevent contamination by foodborne pathogens (Wyrwa & Barska, 2017). The common properties of active packaging material include absorbing and scavenging systems, temperature regulator, potent antioxidant and increasing pathogen susceptibility (Prasad & Kochhar, 2014). On the other hand, intelligent packaging is another concept where the enclosed food material is constantly monitored to ensure quality. Types of intelligent packaging may include those that monitor environmental conditions, quality characteristics or data carriers (Müller & Schmid, 2019). Sensors and indicators coupled with intelligent packaging include time-temperature, calorimetric, biometric, optical, humidity and gas sensors. Some of the techniques in intelligent packaging include barcoding, radio frequency identification and deployment of sensors (Firouz et al., 2021). As a whole, smart packaging comprises techniques like active and intelligent packaging, as shown in Figure 5.5, which enhance consumer interaction with the packaging and food material in a viable manner, promoting the functional value of the packaging material (Young et al., 2020).

These smart packaging techniques could aid in meeting demands incurred in the food chain by ensuring convenience and benefits to the consumer (Biji et al., 2015). Recently biopolymers have been employed to fabricate smart packaging systems to replace conventional packaging materials with sustainable, biodegradable ones (Sani et al., 2021). The biocomposites explored for use in smart packaging material are listed in Table 5.1.

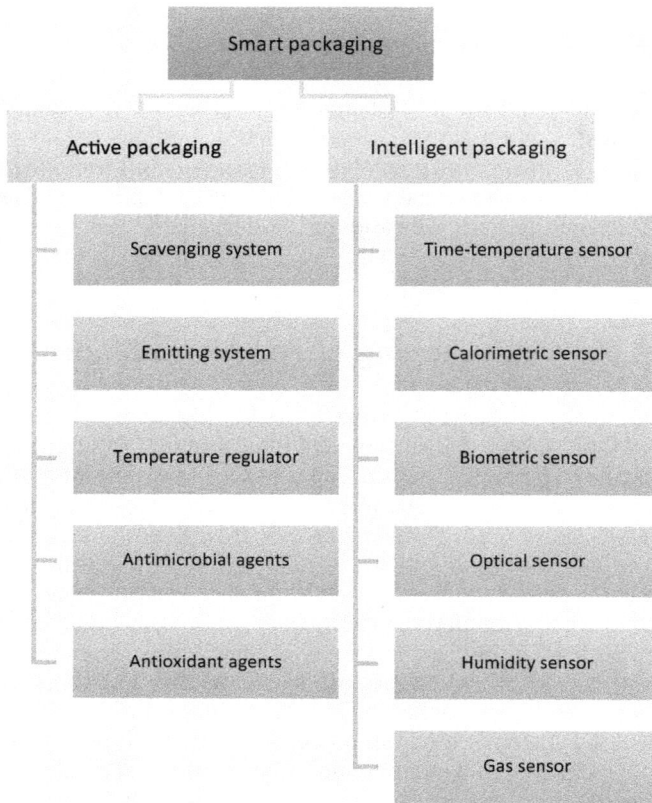

Figure 5.5 Types of smart packaging materials.

5.9 Conclusion

The need for sustainable solutions for food packaging has sharply, increased leading to the exploration of all possible biodegradable materials. However, in order to improve the properties and applicability of biopolymers, they can be combined with other polymers as biocomposites with improved properties to ensure good handling characteristics for the consumer. Thus, this chapter outlines the types of biocomposites used recently for the development of packaging material, along with their position in smart packaging techniques.

Table 5.1 Biocomposites in Smart Packaging Material

Biocomposite	Use	Reference
Chitosan and kombucha tea	Film coating of minced beef	(Ashrafi et al., 2018)
Cellulose acetate and silver nanoparticles	Antimicrobial activity	(Dairi et al., 2019)
Chitosan, beta cyclodextrin citrate with clove oil	Antioxidant and antimicrobial activity	(Adel et al., 2019)
Pea starch, natural antimicrobial agents	Increased shelf life	(Saberi et al., 2017)
Polylactic acid and thymol, carvacrol, limonene, cinnamaldehyde	Antioxidant activity	(Siddiqui et al., 2021)
Polyvinyl alcohol and deacetylated crab shell	Antibacterial activity	(Liu et al., 2022)
Cassava starch and mango pulp	Antioxidant activity	(Reis et al., 2015)
Polylactic acid with silver nanoparticles	Antimicrobial activity	(Fortunati et al., 2013)
Cassava starch and cinnamaldehyde	Antimicrobial activity	(de Souza et al., 2014)
Chitin	Extended shelf life	(Râpa & Vasile, 2020)
Cassava starch and gelatin	Antioxidant activity	(Tongdeesoontorn et al., 2021)
Starch and chitosan	Antifungal activity	(Moeini et al., 2020)

References

Abdullah, N. A., Mohamad, Z., Khan, Z. I., Jusoh, M., Zakaria, Z. Y., & Ngadi, N. (2021). Alginate based sustainable films and composites for packaging: A review. *Chemical Engineering*, 83.

Adel, A. M., Ibrahim, A. A., El-Shafei, A. M., & Al-Shemy, M. T. (2019). Inclusion complex of clove oil with chitosan/β-cyclodextrin citrate/ oxidized nanocellulose biocomposite for active food packaging. *Food Packaging and Shelf Life*, 20, 100307.

Alamri, M. S., Qasem, A. A., Mohamed, A. A., Hussain, S., Ibraheem, M. A., Shamlan, G., . . . Qasha, A. S. (2021). Food packaging's materials: A food safety perspective. *Saudi Journal of Biological Sciences*, 28(8), 4490–4499.

Alves, V. D., Castelló, R., Ferreira, A. R., Costa, N., Fonseca, I. M., & Coelhoso, I. M. (2011). Barrier properties of carrageenan/pectin bio-degradable composite films. *Procedia Food Science*, 1, 240–245.

Angellier-Coussy, H., Guillard, V., Guillaume, C., & Gontard, N. (2011). Wheat gluten (WG)-based materials for food packaging. In *Multifunctional and Nanoreinforced Polymers for Food Packaging* (pp. 649–668). Woodhead Publishing.

Arcan, İ., Boyacı, D., & Yemenicioğlu, A. (2017). The use of zein and its edible films for the development of food packaging materials. *Module in Food Science*. https://doi.org/10.1016/B978-0-08-100596-5.21126-8

Asgher, M., Qamar, S. A., Bilal, M., & Iqbal, H. M. (2020). Bio-based active food packaging materials: Sustainable alternative to conventional petrochemical-based packaging materials. *Food Research International*, 137, 109625.

Ashrafi, A., Jokar, M., & Nafchi, A. M. (2018). Preparation and characterization of biocomposite film based on chitosan and kombucha tea as active food packaging. *International Journal of Biological Macromolecules*, 108, 444–454.

Aydin, F., Kahve, H. I., & Ardic, M. (2017). Lipid-based edible films. *Journal of Scientific and Engineering Research*, 4(9), 86–92.

Bayer, I. S. (2021). Zein in food packaging. *Sustainable Food Packaging Technology*, 199–224.

Berk, Z. (2013). Food packaging. *Food Process Engineering and Technology*, 621–636.

Berk, Z. (2018). *Food Process Engineering and Technology*. Academic Press.

Bhat, K. M., Jyothsana, R., Sharma, A., & Rao, N. N. (2020). Carrageenan-based edible biodegradable food packaging: A review. *International Journal of Food Science and Nutrition*, 5(4), 69–75.

Biji, K. B., Ravishankar, C. N., Mohan, C. O., & Srinivasa Gopal, T. K. (2015). Smart packaging systems for food applications: A review. *Journal of Food Science and Technology*, 52(10), 6125–6135.

Bonnaillie, L. M., Zhang, H., Akkurt, S., Yam, K. L., & Tomasula, P. M. (2014). Casein films: The effects of formulation, environmental conditions and the addition of citric pectin on the structure and mechanical properties. *Polymers*, 6(7), 2018–2036.

Bueno, J. N., Corradini, E., de Souza, P. R., Marques, V. D. S., Radovanovic, E., & Muniz, E. C. (2021). Films based on mixtures of zein, chitosan, and PVA: Development with perspectives for food packaging application. *Polymer Testing*, 101, 107279.

Bugnicourt, E., Schmid, M., Nerney, O. M., Wildner, J., Smykala, L., Lazzeri, A., & Cinelli, P. (2013). Processing and validation of whey-protein-coated films and laminates at semi-industrial scale as novel recyclable food packaging materials with excellent barrier properties. *Advances in Materials Science and Engineering*, 2013.

Chen, H., Wang, J., Cheng, Y., Wang, C., Liu, H., Bian, H., . . . Han, W. (2019). Application of protein-based films and coatings for food packaging: A review. *Polymers*, 11(12), 2039.

Dairi, N., Ferfera-Harrar, H., Ramos, M., & Garrigós, M. C. (2019). Cellulose acetate/AgNPs-organoclay and/or thymol nano-biocomposite films with combined antimicrobial/antioxidant properties for active food packaging use. *International Journal of Biological Macromolecules*, 121, 508–523.

Daniloski, D., Petkoska, A. T., Lee, N. A., Bekhit, A. E. D., Carne, A., Vaskoska, R., & Vasiljevic, T. (2021). Active edible packaging based on milk proteins: A route to carry and deliver nutraceuticals. *Trends in Food Science & Technology*, 111, 688–705.

de Souza, A. C., Dias, A. M., Sousa, H. C., & Tadini, C. C. (2014). Impregnation of cinnamaldehyde into cassava starch biocomposite films using supercritical fluid technology for the development of food active packaging. *Carbohydrate Polymers*, 102, 830–837.

Firouz, M. S., Mohi-Alden, K., & Omid, M. (2021). A critical review on intelligent and active packaging in the food industry: Research and development. *Food Research International*, 141, 110113.

Flórez, M., Guerra-Rodríguez, E., Cazón, P., & Vázquez, M. (2022). Chitosan for food packaging: Recent advances in active and intelligent films. *Food Hydrocolloids*, 124, 107328.

Fortunati, E., Peltzer, M., Armentano, I., Jiménez, A., & Kenny, J. M. (2013). Combined effects of cellulose nanocrystals and silver nanoparticles on the barrier and migration properties of PLA nano-biocomposites. *Journal of Food Engineering*, 118(1), 117–124.

Frank, K., Garcia, C. V., Shin, G. H., & Kim, J. T. (2018). Alginate biocomposite films incorporated with cinnamon essential oil nanoemulsions: Physical, mechanical, and antibacterial properties. *International Journal of Polymer Science*, 2018.

García-Guzmán, L., Cabrera-Barjas, G., Soria-Hernández, C. G., Castaño, J., Guadarrama-Lezama, A. Y., & Rodríguez Llamazares, S. (2022). Progress in starch-based materials for food packaging applications. *Polysaccharides*, 3(1), 136–177.

González, A., & Igarzabal, C. I. A. (2013). Soy protein–Poly (lactic acid) bilayer films as biodegradable material for active food packaging. *Food Hydrocolloids*, 33(2), 289–296.

Gutiérrez, T. J., Mendieta, J. R., & Ortega-Toro, R. (2021). In-depth study from gluten/PCL-based food packaging films obtained under reactive extrusion conditions using chrome octanoate as a potential food grade catalyst. *Food Hydrocolloids*, 111, 106255.

Hajji, S., Kchaou, H., Bkhairia, I., Salem, R. B. S. B., Boufi, S., Debeaufort, F., & Nasri, M. (2021). Conception of active food packaging films based on crab chitosan and gelatin enriched with crustacean protein hydrolysates with improved functional and biological properties. *Food Hydrocolloids*, 116, 106639.

Halonen, N., Pálvölgyi, P. S., Bassani, A., Fiorentini, C., Nair, R., Spigno, G., & Kordas, K. (2020). Bio-based smart materials for food packaging and sensors–a review. *Frontiers in Materials*, 7, 82.

Han, J. W., Ruiz-Garcia, L., Qian, J. P., & Yang, X. T. (2018). Food packaging: A comprehensive review and future trends. *Comprehensive Reviews in Food Science and Food Safety*, 17(4), 860–877.

Huang, X., Zhou, X., Dai, Q., & Qin, Z. (2021). Antibacterial, antioxidation, UV-blocking, and biodegradable soy protein isolate food packaging film with mangosteen peel extract and ZnO nanoparticles. *Nanomaterials*, 11(12), 3337.

Kandasamy, S., Yoo, J., Yun, J., Kang, H. B., Seol, K. H., Kim, H. W., & Ham, J. S. (2021). Application of whey protein-based edible films and coatings in food industries: An updated overview. *Coatings*, 11(9), 1056.

Khan, M. R., Sadiq, M. B., & Mehmood, Z. (2020). Development of edible gelatin composite films enriched with polyphenol loaded nanoemulsions as chicken meat packaging material. *CyTA-Journal of Food*, 18(1), 137–146.

Khan, M. R., Volpe, S., Valentino, M., Miele, N. A., Cavella, S., & Torrieri, E. (2021). Active casein coatings and films for perishable foods: Structural properties and shelf-life extension. *Coatings*, 11(8), 899.

Kraśniewska, K., Pobiega, K., & Gniewosz, M. (2019). Pullulan–biopolymer with potential for use as food packaging. *International Journal of Food Engineering*, 15(9).

Kumar, A., Rao, K. M., & Han, S. S. (2018). Application of xanthan gum as polysaccharide in tissue engineering: A review. *Carbohydrate Polymers*, 180, 128–144.

Lim, L. T. (2015). Enzymes for food-packaging applications. In *Improving and Tailoring Enzymes for Food Quality and Functionality* (pp. 161–178). Woodhead Publishing.

Liu, J., Xu, J., Chen, Q., Ren, J., Wang, H., & Kong, B. (2022). Fabrication and characterisation of poly (vinyl alcohol)/deacetylated crab-shell particles biocomposites with excellent thermomechanical and antibacterial properties as active food packaging material. *Food Biophysics*, 1–11.

Lu, Y., Luo, Q., Chu, Y., Tao, N., Deng, S., Wang, L., & Li, L. (2022). Application of gelatin in food packaging: A review. *Polymers*, 14(3), 436.

Luís, Â., Ramos, A., & Domingues, F. (2021). Pullulan–apple fiber biocomposite films: Optical, mechanical, barrier, antioxidant and antibacterial properties. *Polymers*, 13(6), 870.

Malhotra, S. P. K., & Alghuthaymi, M. A. (2022). Biomolecule-assisted biogenic synthesis of metallic nanoparticles. *Agri-Waste and Microbes for Production of Sustainable Nanomaterials*, 139–163.

Mallick, N., Pattanayak, D. S., Soni, A. B., & Pal, D. (2020). Starch based polymeric composite for food packaging applications. *Journal Engineering Research and Application*, 10, 11–34.

Marsh, K., & Bugusu, B. (2007). Food packaging—roles, materials, and environmental issues. *Journal of Food Science*, 72(3), R39–R55.

Meydanju, N., Pirsa, S., & Farzi, J. (2022). Biodegradable film based on lemon peel powder containing xanthan gum and TiO_2–Ag nanoparticles: Investigation of physicochemical and antibacterial properties. *Polymer Testing*, 106, 107445.

Moeini, A., Mallardo, S., Cimmino, A., Dal Poggetto, G., Masi, M., Di Biase, M., . . . Santagata, G. (2020). Thermoplastic starch and bioactive chitosan sub-microparticle biocomposites: Antifungal and chemico-physical properties of the films. *Carbohydrate Polymers*, 230, 115627.

Mohamed, S. A., El-Sakhawy, M., & El-Sakhawy, M. A. M. (2020). Polysaccharides, protein and lipid-based natural edible films in food packaging: A review. *Carbohydrate Polymers*, 238, 116178.

Mohsin, A., Zaman, W. Q., Guo, M., Ahmed, W., Khan, I. M., Niazi, S., . . . Zhuang, Y. (2020). Xanthan-Curdlan nexus for synthesizing edible food packaging films. *International Journal of Biological Macromolecules*, 162, 43–49.

Müller, P., & Schmid, M. (2019). Intelligent packaging in the food sector: A brief overview. *Foods*, 8(1), 16.

Mustapha, F. A., Jai, J., Sharif, Z. I. M., & Yusof, N. M. (2019, April). Cassava starch/carboxymethylcellulose biocomposite film for food paper packaging incorporated with turmeric oil. In *IOP Conference Series: Materials Science and Engineering* (Vol. 507, No. 1, p. 012008). IOP Publishing.

Mutmainna, I., Tahir, D., Gareso, P. L., Ilyas, S., & Saludung, A. (2019, August). Improving degradation ability of composite starch/chitosan by additional pineapple leaf microfibers for food packaging applications. In *IOP Conference Series: Materials Science and Engineering* (Vol. 593, No. 1, p. 012024). IOP Publishing.

Nilsuwan, K., Arnold, M., Benjakul, S., Prodpran, T., & de la Caba, K. (2021). Properties of chicken protein isolate/fish gelatin blend film incorporated with phenolic compounds and its application as pouch for packing chicken skin oil. *Food Packaging and Shelf Life*, 30, 100761.

Onyeaka, H., Obileke, K., Makaka, G., & Nwokolo, N. (2022). Current research and applications of starch-based biodegradable films for food packaging. *Polymers*, 14(6), 1126.

Oymaci, P., & Altinkaya, S. A. (2016). Improvement of barrier and mechanical properties of whey protein isolate based food packaging films by incorporation of zein nanoparticles as a novel bionanocomposite. *Food Hydrocolloids*, 54, 1–9.

Paine, F. A., & Paine, H. Y. (2012). *A Handbook of Food Packaging*. Springer Science & Business Media.

Patni, N., Yadava, P., Agarwal, A., & Maroo, V. (2014). An overview on the role of wheat gluten as a viable substitute for biodegradable plastics. *Reviews in Chemical Engineering*, 30(4), 421–430.

Pena Serna, C., & Lopes Filho, J. F. (2015). Biodegradable zein-based blend films: Structural, mechanical and barrier properties. *Food Technology and Biotechnology*, 53(3), 348–353.

Porta, R., Sabbah, M., & Di Pierro, P. (2022). Bio-based materials for packaging. *International Journal of Molecular Sciences*, 23(7), 3611.

Prasad, P., & Kochhar, A. (2014). Active packaging in food industry: A review. *Journal of Environmental Science, Toxicology and Food Technology*, 8(5), 1–7.

Praseptiangga, D., Widyaastuti, D., Panatarani, C., & Joni, I. M. (2021). Development and characterization of semi-refined iota carrageenan/ SiO2-ZnO bionanocomposite film with the addition of cassava starch for application on minced chicken meat packaging. *Foods*, 10(11), 2776.

Prusty, K., Patra, S., & Swain, S. K. (2021). Soy protein based biocomposites as ideal packaging materials. In *Biopolymers and Biocomposites from Agro-Waste for Packaging Applications* (pp. 65–84). Woodhead Publishing.

Ramos, M., Valdés, A., Beltran, A., & Garrigós, M. C. (2016). Gelatin-based films and coatings for food packaging applications. *Coatings*, 6(4), 41.

Râpa, M., & Vasile, C. (2020). Processing and properties of chitosan and/or chitin biocomposites for food packaging. In *Food Packaging* (pp. 291–326). CRC Press.

Reichert, C. L., Bugnicourt, E., Coltelli, M. B., Cinelli, P., Lazzeri, A., Canesi, I., . . . Schmid, M. (2020). Bio-based packaging: Materials, modifications, industrial applications and sustainability. *Polymers*, 12(7), 1558.

Reis, L. C. B., de Souza, C. O., da Silva, J. B. A., Martins, A. C., Nunes, I. L., & Druzian, J. I. (2015). Active biocomposites of cassava starch: The effect of yerba mate extract and mango pulp as antioxidant additives on the properties and the stability of a packaged product. *Food and Bioproducts Processing*, 94, 382–391.

Risch, S. J. (2009). Food packaging history and innovations. *Journal of Agricultural and Food Chemistry*, 57(18), 8089–8092.

Roy, S., Priyadarshi, R., & Rhim, J. W. (2021). Development of multi-functional pullulan/chitosan-based composite films reinforced with ZnO nanoparticles and propolis for meat packaging applications. *Foods*, 10(11), 2789.

Rudin, A., & Choi, P. (2012). *The Elements of Polymer Science and Engineering*. Academic Press.

Saberi, B., Chockchaisawasdee, S., Golding, J. B., Scarlett, C. J., & Stathopoulos, C. E. (2017). Characterization of pea starch-guar gum biocomposite edible films enriched by natural antimicrobial agents for active food packaging. *Food and Bioproducts Processing*, 105, 51–63.

Saez-Orviz, S., Laca, A., Rendueles, M., & Díaz, M. (2017). Approaches for casein film uses in food stuff packaging. *Afinidad*, 74(577).

Sanchez-García, M. D. (2011). Carrageenan polysaccharides for food packaging. In *Multifunctional and Nanoreinforced Polymers for Food Packaging* (pp. 594–609). Woodhead Publishing.

Sani, M. A., Azizi-Lalabadi, M., Tavassoli, M., Mohammadi, K., & McClements, D. J. (2021). Recent advances in the development of smart and active biodegradable packaging materials. *Nanomaterials*, 11(5), 1331.

Schmid, M., Merzbacher, S., Brzoska, N., Müller, K., & Jesdinszki, M. (2017). Improvement of food packaging-related properties of whey protein isolate-based nanocomposite films and coatings by addition of montmorillonite nanoplatelets. *Frontiers in Materials*, 4, 35.

Senturk Parreidt, T., Müller, K., & Schmid, M. (2018). Alginate-based edible films and coatings for food packaging applications. *Foods*, 7(10), 170.

Siddiqui, M. N., Redhwi, H. H., Tsagkalias, I., Vouvoudi, E. C., & Achilias, D. S. (2021). Development of bio-composites with enhanced antioxidant activity based on poly (lactic acid) with thymol, carvacrol, limonene, or cinnamaldehyde for active food packaging. *Polymers*, 13(21), 3652.

Souza, V. G., Pires, J. R., Rodrigues, C., Coelhoso, I. M., & Fernando, A. L. (2020). Chitosan composites in packaging industry—current trends and future challenges. *Polymers*, 12(2), 417.

Tongdeesoontorn, W., Mauer, L. J., Wongruong, S., Sriburi, P., Reungsang, A., & Rachtanapun, P. (2021). Antioxidant films from cassava starch/ gelatin biocomposite fortified with quercetin and TBHQ and their applications in food models. *Polymers*, 13(7), 1117.

Vishnuvarthanan, M., & Rajeswari, N. (2019). Preparation and characterization of carrageenan/silver nanoparticles/Laponite nanocomposite coating on oxygen plasma surface modified polypropylene for food packaging. *Journal of Food Science and Technology*, 56(5), 2545–2552.

Wahab, I. F., & Abd Razak, S. I. (2016). Polysaccharides as composite biomaterials. *Composites from Renewable and Sustainable Materials*, 65–84.

Wang, H., Qian, J., & Ding, F. (2018). Emerging chitosan-based films for food packaging applications. *Journal of Agricultural and Food Chemistry*, 66(2), 395–413.

Wang, J., Euring, M., Ostendorf, K., & Zhang, K. (2021). Biobased materials for food packaging. *Journal of Bioresources and Bioproducts*, 7(1), 1–13.

Wyrwa, J., & Barska, A. (2017). Innovations in the food packaging market: Active packaging. *European Food Research and Technology*, 243(10), 1681–1692.

Young, E., Mirosa, M., & Bremer, P. (2020). A systematic review of consumer perceptions of smart packaging technologies for food. *Frontiers in Sustainable Food Systems*, 4, 63.

Yousuf, B., Sun, Y., & Wu, S. (2021). Lipid and lipid-containing composite edible coatings and films. *Food Reviews International*, 1–24.

Yu, Z., Dhital, R., Wang, W., Sun, L., Zeng, W., Mustapha, A., & Lin, M. (2019). Development of multifunctional nanocomposites containing cellulose nanofibrils and soy proteins as food packaging materials. *Food Packaging and Shelf Life*, 21, 100366.

Zafar, R., Zia, K. M., Tabasum, S., Jabeen, F., Noreen, A., & Zuber, M. (2016). Polysaccharide based bionanocomposites, properties and applications:

A review. *International Journal of Biological Macromolecules*, 92, 1012–1024.

Žiūkaitė, A., Strykaitė, M., & Damašius, J. (2022). Screening of cellulose/alginate biocomposites for waterproof food packaging. *Journal of Natural Fibers*, 1–12.

Chapter 6

Biospeckle Laser Technique—A Novel Non-Destructive Approach for Quality Detection

Randeep Kaur, Pooja Nikhanj, and Swati Kapoor

6.1 Introduction

Agricultural products, like vegetables and fruits, are the most important suppliers of nutrients for human health. However, these agricultural products are perishable, have a short shelf life and can quickly decline in quality during the processes of handling, processing and transportation; thus, ensuring their safety and quality from farm to table is vital. Fresh produce spoils and goes down in quality as a result of physiological, biological, biochemical and microbiological changes (Ramos et al., 2013). Changes in texture, moisture content, wilting, crushing and bruising are some of the physical elements that lead to degradation. Some biochemical changes, on the other hand, are caused by enzymatic or non-enzymatic alterations, as well as oxidation processes. Pest assault and deterioration by microorganisms, on the other hand, play a critical part in food spoilage. Therefore, prior

DOI: 10.1201/9781003217138-6

to distribution in the market, fresh products must undergo quality testing to avoid economic losses. Quality testing should be carried out based on the inner and exterior factors that cause produce to deteriorate. With regard to vegetable and fruit size, colour, shape, texture and appearance are external factors, whereas titratable acidity, pH, soluble solids, starch and sugar content, ratio of soluble solids to titratable acidity, total phenolic compounds, carotenoids, total antioxidants and enzymatic actions are interior factors (Mesa et al., 2016) contributing to product quality.

Inadequate storage conditions and incorrect management can bodily harm the product, and unwanted biochemical changes render it unsuitable for ingestion. Quality evaluation is critical during the post-harvest and preharvest periods. The preservation procedure is difficult due to the rapid degradation and short shelf period of fresh fruits and vegetables. Because of quick multiplication of microorganisms, spoiled product in a pack may result in further fast decay of other things associated with the produce. As a result, farmers are being encouraged to employ blended insecticides or other similar chemicals in agricultural goods to extend their shelf life and postpone ageing. Toxic synthetic pesticides are commonly utilized in the Indian farming business, putting customers' health at risk (USEPA, 2015). As a result, inspecting the quality of fresh products, particularly fruits and vegetables, before consumption is critical. Because of the potency, minimum effort, high accuracy and simplicity, optical methods have been shown to be critical and effective in quality assurance. The ability to report on the state of the microbial load and quality parameters in near real time is a fundamental feature of an ideal monitoring system. As a result of the time delay in the microbial culture procedures, traditional techniques for measuring microbial load in agro-products are ineffective (Murray et al., 2017). The methods traditionally used for quality detection of fresh produce are destructive, consume more time and require more labour in spite of this and require a lot of chemicals, thus being uneconomical. Chromatographic technologies, such as gas liquid chromatography (GLC) and high performance liquid chromatography (HPLC), with high analytical costs and a destructive nature are restricted in everyday situations to test food product quality (Lee et al., 2015).

Therefore, an ocular technique for detection of quality in the food industry is needed. Currently, non-destructive techniques for monitoring the safety and quality of food commodities are drawing attention due to their ease of process and economy. Based upon the optical, electrical and ultrasonic characteristic principles, quality evaluation equipment has been designed to ensure the quality and safety of fresh products. Non-destructive quality-finding processes are quick, simple and cost-effective, ensuring the quality and safety of fresh products such as fruits and vegetables. Microscopy, functionalized atomic force microscopy, laser scattering, hyperspectral imaging, vibrational spectroscopic techniques and Doppler spectroscopy are some of the optical technologies used for non-destructive food analysis. For good results, microscopic inspection requires constant practice and significant training, but vibrational spectroscopic techniques are less vulnerable since they require a larger calibration set for monitoring food quality.

On other hand, hyperspectra imaging techniques cannot screen the entire exterior area of fruits and vegetables; it results in lower accuracy of the technique. Because of the inadequacies of these techniques, they cannot distinguish the various defects such as disease, scabs, bruising and fungal infection in fruits and vegetables. To overcome these inconveniences, researchers have been working on the optical phenomenon known as biospeckle laser technique (BSL).

The first time this technique was demonstrated was by Briers in 1975, followed by Asakura in 1988 and Fujii *et al.* in 1985 for monitoring blood flow. The technique was applied in many biological areas, such as analysis of seed viability, maturation and bruising in fruits and vegetables and activity of parasites in living tissues.

6.2 Biospeckle Laser Technique

The biospeckle laser, also known as the dynamic laser speckle (DLS), is a non-destructive and non-invasive method of analysing biological materials. When a biological substance is exposed to a coherent light source, such as a laser, light scatters, generating an interference pattern on an observational plane that is represented by a granular picture known as a speckle pattern. The biospeckle laser technique by definition is an optical, non-destructive quality detection tool executed by analysing the biospeckle activity (BA) of the biological sample.

The biological activity of the material includes cytoplasmic streaming, organelle movement maturation and ripening changes, bacterial and fungal infection, enzyme activities and Brownian motion. These biological activities are considered biospeckle activities that can be analysed non-destructively by using the biospeckle method combined with the other processing techniques. More clearly, biospeckle activity (BA) is defined as the change in the biological activity of a biological material. Biospeckle activity depends upon the surface and inner tissues of the sample. The main objective of BSL technique is to detect changes at the cellular level of the food by correlating them with biospeckle activity. Defects in fruits and vegetables therefore could be identified before the symptoms are visible, as changes occurring in biochemical processes result in change of biospeckle activity, which can thus be used as a tool in quality maintenance and safety measures, especially in fruits and vegetables.

6.3 Features of Biospeckle Laser Technique

- **BSL is an optical tool and a non-destructive technique**
  Biospeckle is a non-destructive method for determining the viability of living materials. In the fields of health, microbiology and agriculture, biospeckle has a wide range of uses. The approach has recently seen widespread use in the sector of agriculture for assessing the quality and safety of food items.
- **Rapid and easy to operate**
  It's a low-cost, basic technology that may be used in the food sector to determine the quality of samples without ever having to touch them.
- **Cost effective and guarantees the security of fresh products**
  The key benefits of laser light technology over other optical techniques are that it may represent organically active parts of the test sample by detecting light interference fluctuations and delivering data that is difficult to get using multispectral techniques or human observation.
- **Identifies maturation and ripening changes**
  The main use is the identification of illness and flaws in fresh food and monitoring of the ageing and maturation process. Chemical exchanges occurring in biological materials, as well as other intracellular activities, including organelle movement, cytoplasmic streaming, cell growth and division and Brownian motion, are determined to be the source of biospeckle activity.
- **Detects bruising and fungal growth in fruits and vegetables**
  By evaluating changes in the biospeckle activity of tissues, biospeckle laser technology may identify contamination as well as damage in fruits and vegetables.
- **Detects aging of meat and viability of seed**
  A biospeckle approach, discovered as an excellent instrument for analysing and measuring biological activity of the ageing process of meat, is based on the moment of inertia and absolute value difference and has shown considerable promise for evaluating beef quality, particularly tenderness and colour.

6.3.1 Methodology

Biospeckle laser technique is executed in four main steps.

Ocular Arrangement → Online Modification → Collection of data → Analysis of data

6.3.2 Ocular Arrangement

In ocular arrangement called experiment set up phase, all devices like laser, camera and lens are positioned according to the biological sample placed for the analysis (Figure 6.1).

6.3.3 Online Modification

This includes the ideal position of the sample, distance between camera and object, distance between laser and object, power of laser, angle of incident (between laser and camera) and image size of speckle pattern. These parameters must be adjusted according to the characteristics of the biological sample. Because the majority of agricultural products have rough surfaces, biospeckle equipment employs laser light (with a wavelength greater than 600 nm) to illuminate them. A charge-coupled device (CCD) camera is used to capture biospeckles. The brightness of the CCD camera will be empirically tuned to prevent over-exposed pixels on the histogram image.

6.3.4 Collection of Data

The data of the evaluated sample is collected in the form of speckle patterns. A speckle pattern is created by mutual interference by a set of coherent wave fronts at the same time. In general, there are two factors of a biological sample that are analysed by the speckle pattern. The immobile particles in the sample produce a static pattern, while the mobile elements of the sample produce a changing pattern. While the inorganic components are linked with the static patterns and the mobility degree of the speckle pattern is

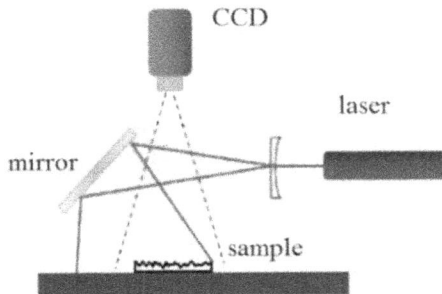

Figure 6.1 Experimental setup used in research with BSL.

defined as by the term "biospeckle activity". Amongst these, biospeckle technique shows variable speckle pattern representing distinguishing feature of biological material (Ansari & Nirala, 2012). The speckle pattern provides information about the changes in biological activity (biochemical process) of a sample. Alteration in the biological activity of a sample results in either an increase or decrease in biospeckle activity.

6.3.5 Analysis of Data

Various methodologies are used to study the strength of biospeckle activity. The activity is primarily assessed using a time-dependent correlation of irradiance. Qualitative and quantitative approaches are the two types of analytical methods. In quantitative analysis, the methods of absolute value difference (AVD), cross-correlation and inertia moment (IM) are frequently used, whereas qualitative analysis frequently uses Fuiji, temporal difference (TD), laser speckle temporal contrast analysis (LASTCA), laser speckle contrast analysis (LASCA) and generalized differences (Retheesh et al., 2018). Different techniques and algorithms to study the intensity of biospeckle are compiled in Table 6.1.

6.4 Application of Biospeckle Laser Technique in Agro-Industry

6.4.1 Detection of Damage in Agro-Produce

Failure of tissue or injury caused by external pressures in forceful manner or motionless settings is a common symptom of mechanical damage in horticultural commodities. At the time of harvest and during post-harvest processes, the fragile skin of fruits and vegetables is prone to injury. Mechanical damage to fresh fruit surfaces serves as a major entrance point for microorganisms, resulting in a loss of quality product and a risk to food safety (Enes et al., 2012). Mechanical damage results in bruising on the fruit, whether visible or not. Due to the action of oxidation reaction of phenolic compounds, tissue browning is the symptom of this damage. Fruit bruises can also form as a consequence of tissues breaking under pressure. Tissue breakdown occurs on a microscopic level, resulting in imperfections of cells beneath the skin without causing any harm to the surface (Pajuelo et al., 2003). The biological activity of injured tissues differs from that of typical healthy tissues. As a result, variations in biological activity may be tracked throughout the whole tissue surface. According to Braga et al. (2007), the generalized difference approach may be used

BIOSPECKLE LASER TECHNIQUE

Table 6.1 Various Algorithms Used for Biospeckle Analysis

Algorithm	Equation	Reference
Absolute value difference	$AVD = \sum_{ij}\{COM[ij]*|i-i|\}$	Braga et al. (2007)
Fujii's method	$F(x,y) = \sum_{K=1}^{N}\frac{|I_k(x,y)-I_{K+1}(x,y)|}{I_k(x,y)+I_{K+1}(x,y)}$	Fujii et al. (1987)
Generalized difference (GD)	$GD(x,y) = \sum_{K=1}^{N-1}\sum_{i=k+1}^{N}|I_k(x,y)-I_i(x,y)|$	Lee et al. (2015)
Inertia moment	$IM = \sum_{ij}M_{ij}(i-i)^2$	Arizaga et al. (1999)
Weighted generalized difference (WGD)	$WGD(w,s) = \sum_{i=0}^{n}\sum_{j=i+1}^{i+w}|x_i-x_i|P_i$	Rabal (2008)
Time history of speckle pattern (THSP)	$\xi = \sqrt{\frac{1}{N-1}\sum_{t=1}^{N-1}(\mu_{t,t+1}-\mu_t)^2}$ μ_i	DaSilva and Muramatsu (2008)
Laser analysis temporal contrast analysis	$K_t(x,y) = \frac{\sqrt{\frac{1}{N-1}\left[\sum_{n=1}^{N}\left[I_{x,y}(n)-I_{x,y_t}\right]^2\right]}}{I_{x,y_t}}$	Pengcheng et al. (2006)
Laser speckle contrast analysis	$K_x = \frac{\sigma_x}{I_x}$	Briers (1978)
Temporal difference	$TD(x,y) = \sum_{K=1}^{N}|I_k(x,y)-I_{K+1}(x,y)|$	Retheesh et al. (2018)
Briers contrast	$\frac{\sigma_t^2(x)xy}{I^2} = 1-(1-\rho)^2$	Briers (1978)
Motion history image (MHI)	$MHI(x,y) = \sum_{J=1}^{n}T_i K_i$	Retheesh et al. (2018)
Autocorrelation	$\gamma(\tau) = \frac{I(t)I(t+\tau)x}{It_x^2}$	Cummins and Swinney (1970)

149

to identify the activity levels of observed speckle pattern changes. The implementation of the biospeckle approach can be used to detect surface and sub-surface defects. The LASCA, WGD and Konishi techniques may all be used to subjectively describe a course of action. The scientists also showed that malformations are more likely to be detected before symptoms appear.

Using intensity-based algorithms like the biospeckle activity value, AVD, IM, GD, co-occurrence matrix, parameterized Fuiji, parameterized generalized difference, grey level co-occurrence matrix (GLCM) and granulometric size dispersion, Kumari and Nirala (2016) conducted a shelf life study on bruised as well as other parts of juicy Indian apple for nine days. The inertia moment approach revealed the difference between clean and bruised region biospeckle activity most effectively of any method tested. The GLCM parameters and BA values were shown to be effective in distinguishing between fresh and damaged Indian apple areas. In order to estimate biospeckle activity intensity, Lee et al. (2015) employed the GD, LASTCA and Fuiji algorithms to estimate the speckle picture. The results showed that biospeckle activities of the undamaged portions of the apple were comparable for all three approaches; however, the damaged region had decreased biospeckle activity after the injury.

Samuel et al. (2017) employed a laser speckle imaging system in conjunction with dispensation algorithms for studying the cross-relationship and history of speckle patterns to evaluate the nature of common organic products like fruits and vegetables, such as mangoes, apples, oranges, guavas and cucumbers. The relationship plotting approach offers a straightforward way to do estimations consistently for a considerable amount of time while avoiding the effects of local aggravations. Utilizing cross-correlation and IM techniques, mechanical damage to organic materials was statistically investigated. The activity maps made using the GD approach successfully identified the affected areas. The technique was also used to identify plant pesticides. The method developed is rapid, useful, and non-contact; it can be modified further to suit a business through robotization, which may benefit ranchers and partners financially. To demonstrate effectiveness of the method, Retheesh et al. (2018) utilized motion history image to examine the scar region surrounding green orange fruit and compare it with other intensity-based methodologies. Researchers believe that MHI might be a very useful online tool for finding flaws.

Sutton and Punja (2017) suggested a unique whole seed evaluation approach for detecting sprouting damage in wheat seeds by combining qualitative and quantitative data from biospeckle estimate measures. The investigators looked at divided seeds and observed that the germ had a high incidence of biospeckle activity, which is associated with germination.

Biospeckle laser analysis can identify emerging damaged kernels, according to the findings of the study. Biospeckle laser analysis was also shown to be more useful in analysing seed germination traits and dormancy processes in wheat varieties. Scientists employed solid-state lasers (532 nm) as a biospeckle source rather than red lasers (633 nm) since they are subject to photo-regulatory effects at 633 nm. In comparison to other spectra, Budagovskii et al. (2012) found that the 532-nm laser wavelength had a minor influence on blackberry root development. Arefi et al. (2016) utilized biospeckle imaging technology to create a model for distinguishing mealy and non-mealy apples wherein 540 apple samples that had been kept in the refrigerator for zero to five months were used for the investigation. Additionally, 220 additional apples were stored in 95% relative humidity at 20°C for 10 to 26 days. Every apple that was stored was biospeckled at a wavelength of 680 and 780 nm. To evaluate the fruit's juiciness and stiffness, an immediate compression test was performed after biospeckle imaging. Following the categorization of apples into categories according to speckle patterns, destructive processes and time-historical studies were created and biospeckle pattern analysis was performed utilizing IM, AVD and autocorrelation approaches. New tasters had higher biospeckle activity than semi-mealy samples and mealy apples. For fresh (81.7%) and semi-mealy (70.9%) apples, the wavelength of 780 nm offered the highest categorization precision. The best result of 77.3% accuracy was found at a wavelength of 680 nm for mealy apples. The researchers concluded that the biospeckle approach might be used to identify mealiness in apple fruit in a non-destructive manner.

Abou-Nader et al. (2019) used a basic optical setup to explore the light scattering characteristics of apple flesh. Despite the fact that the apples were of unknown origin, had been stored before and had a shorter shelf life, a link was discovered between the fast recorded activity and physiological characteristics of apple samples. These findings revealed a strong link between high-speed inner activity and apple tissue hardness and sugar levels; therefore, this study adds to our understanding of the important time scales for apple optical examination. Gao and Rao (2019) performed research on black spot bruise incidence and examined its biospeckle imaging responsiveness activity in potatoes. Two enhancements to the method were suggested in this work: correlation of contiguous speckle pattern wavelet entropy (CCSP WE) and correlation of contiguous speckle pattern–modified wavelet entropy (CCSP MWE). The number of potato samples had no influence on the frequency of the black spot bruises, but the stem end of the potato was more bruise susceptible, according to the susceptibility study. As a result, the CCSP WE approach was shown to have a higher weight to volume ratio and was effective in detecting black spot bruises.

6.4.2 Detection of Aging

6.4.2.1 Aging Detection in Meat

According to Isis et al. (2013), the evaluation and measurement of the biological processes of beef ageing was explored using the laser biospeckle technology and image processing. Biospeckle laser restrictions can be used to assess the biological activity brought on by the actions of the endogenous enzymes calpains and cathepsins, which are in charge of the ageing processes in cattle. It is based upon the relationship between an examination of Warner–Bratzler shear force (WBSF) and enzyme activity, a measure of meat softness that correlates with time spent maturing. The biospeckle method with IM and absolute value difference was found to be a useful tool for analysis and quantification of biological activity during the ageing process of meat. This method possesses great potential for evaluating the quality of beef, particularly tenderness and colour.

According to Oleksandr et al. (2015), the modified spatial temporal speckle association technique was used to quantify post-slaughter muscle changes in an efficient manner. Changes regarding biochemical processes in tissue were measured by biological activity coefficient. It has been verified that the biological activity coefficient may be used to assess the activity of tissue samples from pigs and chickens muscles. According to the spatial temporal dynamics analysis of biospeckle patterns, the bioactivity level was observed to decline during the long-term storage period as shown by the rupturing and cracking of muscle fibres, the shrinkage of sarcomeres, nuclei deformation, the diminution of nuclear chromatin and the destruction of mitochondria.

6.4.2.2 Age Detection in Fruits and Vegetables

Lasio and Zude (2009) used backscattering imaging and the Monte Carlo (MC) approach to conduct a biospeckle investigation of kiwi fruit tissue. To reduce the amount of photons required, a time-resolved Monte Carlo model was applied. Rotation in the intensity profiles of backscattering pictures of kiwi fruit was observed due to changed anisotropy. By comparing the measured intensity patterns, the fruits' anisotropy factor was estimated. In terms of textural qualities, there was a substantial difference between over-ripe and premium-quality fruit anisotropy. Ansari and Nirala (2012) evaluated biospeckle activity for three dissimilar organic Indian goods, pear, apple and tomato, using two methodologies, spatial temporal speckle correlation and IM. Bioactivity was determined using a biospeckle THSP picture taking the IM and intensity changes cross correlations into consideration. Aside from shelf life changes, bioactivity key modulations were explored, and it was discovered that pear had more activity than tomato and apple, indicating that the IM method was successful. Furthermore, because the biospeckle activity of organic products decreases with maturity, pear exhibited a higher drop than tomato and apple as predicted by the cross-correlation method's result. These activity variations are thought to be based on their breathing rates. In comparison to the cross-correlation approach, IM is found to be more reliable in predicting organic goods' bioactivity levels, as IM assesses bioactivity specifically.

In order to determine the age of lemons, Ansari and Nirala (2016) used a biospeckle activity with dynamic speckle pattern technique. The moment of inertia and the spatial temporal speckle correlation coefficient were used to quantify and analyse the speckle pattern of lemon fruits. THSP line profiles were used to characterize biospeckle activity. The authors discovered that a dynamic speckle measure changes when the quality of the fruit declines and that measures vary as the picture of the fruit moves. Retheesh et al. (2018) employed the biospeckle approach to investigate the ripening of mango fruit and proved its efficacy. The scientists assessed ripening changes and forecasted the fruit's shelf life using time history of speckle pattern and 2D cross-correlation function analysis. The amount of moisture in the fruits, the speed with which images are captured, the optimal temperature, the testing conditions and so on all have an impact on the final results. As a result, the authors recommend that these variables be kept constant throughout the experiment. Costa et al. (2017) used biospeckle laser technology to investigate the maturation phase of macaw palm fruit. The authors discovered a link between biospeckles, which assess biological activity, and the hardness of the macaw palm fruit pulp, which is typically picked depending on soil composition. This discovery had a significant impact on when the fruit should be harvested.

Piotr et al. (2018) created a biospeckle approach for evaluating and monitoring tomato ripening stages. For collecting biospeckle patterns, laser light with wavelengths of 830 and 640 nm was used. The de-correlation patterns in biospeckles were examined using C4 and activity descriptors. As a reference approach, the optical properties of tomato skin were employed, including chroma, luminosity and a^*/b^* ratio. The techniques were tested for their ability to predict tomato growth and damaging indicators such as firmness, chlorophyll and carotenoid concentration, among other things. For predicting fruit firmness, chlorophyll and carotenoids, the laser wavelength of 640 nm was revealed to be more precise than the laser wavelength of 830 nm, and these characteristics were significantly connected with brightness. Matheus et al. (2017) used a biospeckle approach to analyse the moment of inertia to develop a method to identify the sugar and water content in sugar cane. Internal and exterior husk sample analyses were shown to be compatible with this approach. The authors also used non-destructive and non-invasive NIR spectroscopy and biospeckle methods to evaluate the moisture and sugar content in sugar cane.

6.5 Detection of Microbial Contamination

The use of biospeckle as a technique for investigating processes in microbiological media has progressed significantly. Murialdo et al. (2009) used biospeckle experiments to demonstrate that *Pseudomonas aeruginosa* shows chemotaxis on agar media plates, as well as to distinguish fungi from other motile bacteria present in the same culture (Murialdo et al., 2009). Biospeckle analysis was used to assess *E. coli* movement in solid media, and the temporal difference approach was used. González-Peña et al. (2014) used speckle patterns for examining cell malignancy and drug sensitivity. Additionally, Vladimirov et al. (2016) looked into the intracellular activities of certain cells that had been exposed to coherent light in order to connect the activity of these biological activities to the effects of on frames of speckles. Enes et al. (2012) used the biospeckle approach in conjunction with digital image information technology (DIT) to examine bacterial growth and identify medication effects on *Trypanosoma cruzi*, a parasitic euglenoid. A parameterized temporal difference approach was used to quantify the activity. The method was proven effective in observing morphological changes and motility in bacterial populations over time, as well as identifying and distinguishing short-term medication action on parasites. As a result, the optical imaging system is a valuable tool in the food microbiology and biotechnology industries for determining pathogen infection in foods swiftly.

Adamiak et al. (2012) employed the biospeckle technique to investigate bull's eye rot disease in apples and quality of product throughout storage under various conditions as well as to approximate the product's shelf life. Equipment of 8 MW, an extended beam and 635-nm lasers was used to light the apples and the equator at six places. The correlation coefficient was calculated using a CCD camera recorder. Biospeckle activity increased with the advent of fungal illness during cold storage, while signs of bull's eye rot disease were barely visible on apple samples. Senescence showed a decrease in biospeckle activity due to a decrease in biological activities inside the tissue. Throughout the shelf life study, there was also a decline in biospeckle activity, regardless of storage conditions.

The biospeckle method was followed by Rabelo et al. (2011) and Braga (2017) for identifying fungal infection in seedlings (2005). Braga et al. (2005) employed IM and GD processes to assess biospeckle variation in bean seeds, in addition to Fuiji's methodologies. The technique's capacity to detect microorganisms in beans was proved in these studies. Seedlings treated with fungus had greater IM values than control seeds (indicating strong biospeckle activity). A certain range of IM values may be used to differentiate two of the three fungal species. The presence of fungus was confirmed using GD and Fuiji methods. For examination of bean seeds treated with *Aspergillus flavus* or *Fusarium oxysporum*, Rabelo et al. (2011) employed spatial time speckle (STS) and IM signal frequency. Using the Fourier transform technique and a He-Ne red beam laser with a CCD camera, STS signal frequencies were generated and recorded. The IM values of infected seeds were found to be greater than those of control seeds. The presence of fungus was confirmed using the Fuji and GD methods. According to the researchers, frequency analysis might be utilized to improve IM values, boosting the usage of biospeckle laser technologies in natural material research.

Biospeckle activity was employed by Jitendra et al. (2017) to design a strategy for screening medicinal plants for microbiological elements. To produce a biospeckle pattern, a laser source was utilized to light fresh and infected *Ficus religiosa* samples. Using an outline grabber, a CCD camera caught dynamic changes in visuals, which were quantized into an 8-bit digital picture. A weighted parameterized Fuji method, an improved version of Fuji's technique, was employed for image analysis. It indicates a high level of complex nature when the test sample is consistently illuminated. The IM technique using biospeckle data was utilized to evaluate the leaf tissue (Ansari & Nirala, 2016). According to the researchers, the IM might be used to display a plant leaf biospeckle signature, and it has been revealed that plants with different leaves have varied IM values. Additional investigation revealed that biospeckle activity in diseased leaf sections was lower than in healthy tissue. This research helped distinguish between fresh and damaged leaves, which are important when making high-quality ayurvedic

Table 6.2 Applications of BSL in Contamination Detection

Sample	Size of Sample, Number of Images	Result	Reference
Bean seeds	–	Fungal infection	Rabelo et al. (2011)
Corn seed	100 samples (480 × 250 pixels)	Viability of seeds	Piotr et al. (2018)
Strawberries	Less than two images per data set	Differentiated between fresh and injured leaves	Jitendra et al. (2017)
Wheat seeds	Sprouting is detected in 256 images over a period of 10 s per sample	Sprouting is detected	Sutton and Punja (2017)

(Continued)

Table 6.2 (Continued)

Sample	Size of Sample, Number of Images	Result	Reference
Orange, guava, mango, apple, cucumber	Dynamic speckle of size 512 × 512	Detection of vegetable scratched areas and pesticides	Samuel et al. (2017)
Peepal leaves	NA	Differentiated fresh and injured leaves	Jitendra et al. (2017)
Green orange	32 images per second of 3.75 × 3.75 μm pixels	Detection of blemish area appropriately.	Retheesh et al. (2018)
Apples	500 images of bruised apple samples (300 pixels) 50 undamaged apples in 512 frames (480 × 320 pixels)	Bruise detection damage Damage area detected has less BA	Lee et al. (2015)
Fresh cut apples	Pixel size of 3.75 × 3.75 μm recorded at an angle of 25°	The degree of polarization and grain speckle image contains physio-chemical information	Minz and Nirala (2016)

medications. Pieczywek et al. (2017) formulated a method for early identification of disease indications in apples infected with the fungus *Pezicula malicorticus*, which causes bull's eye rot, by using new optical sensors to analyse storage changes. The researchers developed a method that detects symptoms in as little as two days using chlorophyll fluorescence, biospeckle activity and hyperspectral image analysis. The typical vision examination took four to five days after the vaccination. When compared to hyperspectral imaging, spatial images of biospeckle activity were shown to be more effective for following sickness development. It has also helped diagnose sickness development considerably in advance when compared to the chlorophyll fluorescence method.

6.6 Effect of Biospeckle Activity on Biochemical Processes

Lasio and Zude (2009) used backscattering imaging and the MC approach to conduct a biospeckle investigation of kiwi fruit tissue. A time-resolved MC model was utilized to minimize the required number of photons. Changes in anisotropy caused alternation in the intensity profiles of backscattering photographs of kiwi fruits. The anisotropy factor of the fruits was compared by observed intensity patterns. A significant variation was observed in textural qualities between overripe and high-grade fruit. The nature of the decline in biospeckle activity as the temperature fell was determined using the Q10 coefficient. The fact that the fruit tissue exhibits a high metabolic activity at a low temperature might be due to metabolic and biochemical processes involving diffusion or Brownian motion (Kurenda et al., 2012). This study used biospeckle activity to reflect the metabolic changes that occur during apple development and maturation.

Biospeckle activity increased during the pre-harvest phase, but soluble solid content increased, while acid and starch content declined somewhat. The researchers discovered a strong link between biospeckle activity, soluble content, starch concentration and hardness, indicating the biospeckle approach has a lot of potential for non-destructive evaluation of pre-harvest characteristics. Biospeckle activity might be used as a new measure of fruit quality, according to this research. Biospeckle activity and apple chlorophyll levels had a linear connection, according to the study. The amount of chlorophyll in the environment has a direct correlation with biospeckle activity. In order to undertake relevant biospeckle research, the right laser wavelengths must be chosen according to the qualities of the substance under study. A laser wavelength of 800–880 nm proved the most efficient for examining intracellular component mobility.

Biochemical changes have a major influence on biospeckle activity in apple postharvest storage, according to studies. When the breakdown of starch granules was studied using a cross-correlation technique, it was observed that apple tissues had fewer scattering centres and extremely poor biospeckle activity. The influence of storage temperature on recorded biospeckle variation was also demonstrated (Kurenda et al., 2012); analyses of indices such as moment of inertia, speckle contrast and correlation coefficient of apple biospeckle dynamics indicated that the indices decreased as the sample cooled. Ansari and Nirala (2012) claimed that monitoring the metabolism and biochemical alterations that take place within the tissues and cells of fruits using biospeckle activity evaluation might be a useful technique. Kurenda et al. (2012) investigated the effects of metabolic inhibitors on intracellular component trafficking, including lantrunculin B, dimethyl sulfoxide (DMSO), colchicine, cytochalasin B and a combination of ion channel inhibitors. Biospeckle activity was examined using the LASCA technique and cross-correlation coefficient. According to the findings, a substantial impact on biospeckle activity was accounted for 74% of rotting in apple tissue with DMSO, latrunculin B and ion channel inhibitors. Furthermore, the findings revealed that actin microfilament and ion channel activities had a significant impact on biospeckle activity. Sutton and Punja (2017) used the biospeckle technique to examine biochemical changes in seed surface movement and determine the viability of the seeds. A 7-mW helium neon laser source of 632.8 nm wavelength was used to light both non-viable and viable *Pisum sativum* seeds. Speckle patterns were recorded by a CCD camera and analysed in a MATLAB. The biological activity of speckle patterns was assessed using GD and IM methods, and it was discovered that the method worked well for identifying non-viable and viable seeds based on biochemical changes. Water activity plays a critical role in the metabolic process, as Silva et al. (2018) showed by demonstrating the link between the two. The scientists showed that biospeckle laser calibration allows for the recording of sizable contrast variations in solutions with a broad range of concentrations in a sizable temperature array. The findings suggest that distinguishing between various forms of water activity in solutions may be accomplished using biospeckle laser technology. Mandracchia et al. (2018) employed biospeckle de-correlation to swiftly assess the effectiveness of microencapsulation for preservation mechanism. It was observed that utilizing biospeckle decorrelation, it was possible to determine probiotic bacteria's survival rate and shelf life under simulated gastrointestinal conditions. The use of an appropriate method for image processing of biospeckle laser activity might improve the quantitative assessment of biological material's physicochemical attributes.

6.7 Limitations and Challenges of Using Biospeckle Techniques

Biospeckle techniques have certain limitations that can be categorized into two main types.

6.7.1 Hardware Limitations

- *Stability of laser*—Measurement errors may occur as a result of the use of low-quality power sources and unstable laser equipment. As a result, diode lasers should only be utilized after performing suitable testing to ensure their stability.

- *Adjustment of camera*—Many capabilities are available in today's cameras, such as zooming, white light balance, speed and so on. Given factors may be automated and set to the default, allowing the camera to respond to changes in lit materials at all times. This may cause changes in lasers' dynamic speckle generation (Braga, 2017). As a result, setting up the camera and stopping automated adjustments is a necessary step before compiling data. The user will be able to employ inexpensive cameras as a result. Furthermore, the f-number refers to the adjustments of the focal lens that adhere to the system accordingly to the diameter of the entry pupil (iris) (Braga, 2017).
- *External noise (portability)*—If the equipment is to be used outside, it should be strengthened to avoid interference caused by mechanical sounds and light from external sources.

6.7.2 Software Limitations

- *Standardization*—Standardizing the methodologies employed in the investigation and comparison of biospeckle phenomena should be done in a well-organized and well-classified manner. Offline and online methodologies are now employed for analysis. The findings of online techniques are produced in graphical representations and include LASCA (Briers, 1978) and MHI. Offline approaches for analysing biospeckle data, on the other hand, can be in the temporal or frequency domain, with graphical or numerical outputs in both cases.
- *Light reference*—Light influence during biospeckle phenomena has to be taken into account, since variations in light intensity might cause changes in the final output.

In order to build a generic method for measuring activity, there is a need for standard and commercial equipment dedicated to agricultural produce. The biospeckle laser's penetration depth on biological samples (at a wavelength of 632.8 nm), however, restricts its use. To illustrate, in apple skin (Lammertyn et al., 2000), the depth of penetration is just 2 mm, but in apple tissue, it is 7–10 mm. Biospeckle activity information may therefore be obtained from tissue beneath the skin. Biospeckle activity in the fruit's core, however, is difficult to acquire. Biospeckle activity is also affected by the surface features of agro-produce (Mulone et al., 2014).

6.8 Conclusion

As studied, biospeckle is a non destructive technique for evaluation of living material. It has extensive applications in the areas of medicine, microbiology and agriculture. It has significant use in assessing the quality and safety of food items. It is a simple, uncomplicated and affordable technique with the potential to be used as a tool for evaluating the quality of samples without actually touching them. There are various obstacles and limitations that must be overcome through the widespread use of techniques other than optical labs.

References

Abou-Nader C, Tualle J M, Tinet E and Ettori D (2019) A new insight into biospeckle activity in apple tissues. *Sensors.* **19**(3): 497.

Adamiak A, Zdunek A, Kurenda A and Rutkowski K (2012) Application of the bio speckle method for monitoring bull's eye rot development and quality changes of apples subjected to various storage methods—preliminary studies. *Sensors.* **12**: 3215–3227.

Ansari M Z and Nirala A K (2012) Biospeckle techniques in quality evaluation of Indian fruits world academy of science. *Eng. Technol.* **6**: 11–20.

Ansari M Z and Nirala A K (2016) Assessment of biospeckle activity of lemon fruit. *Agri. Eng. Int. CIGR J.* **18**: 190–200.

Arizaga R, Trivi M and Rabal H (1999) Speckle time evolution characterization by the co-occurrence matrix analysis. *Opt. Laser Technol.* **31**(2): 163–169.

Braga R A (2017) Challenges to apply the biospeckle laser technique in the field. *Chem. Eng. Trans.* **58**: 577–582.

Braga R A, Horgan G W, Enes A M, Miron D, Rabelo G F and Barreto Filho J B (2007) Biological feature isolation by wavelets in biospeckle laser images. *Comput. Electron. Agric.* **58**: 123–132.

Braga R A, Rabelo G F, Granato L R, Santos E F, Machado J C, Arizaga R and Trivi M (2005) Detection of fungi in beans by the laser biospeckle technique. *Biosyst. Eng.* **91**: 465–469.

Briers J D (1978) The statistics of fluctuating speckle patterns produced by a mixture of moving and stationary scatterers. *Opt. Quantum Electron.* **10**(4): 364–366.

Budagovskii A V, Solovykh N V, Budagovskaya O N, Budagovskii I A, Michtchenko A and Hernandez-Vizuet M (2012) Response of plant organisms to laser irradiation of different spectral composition. *Russ. Agric. Sci.* **38**: 367–370.

Costa A G, Pinto F A, Braga R A, Motoike S Y and Gracia L (2017) Relationship between biospeckle laser technique and firmness of Acrocomiaaculeata fruits. *Rev. Bras. Eng. Agric. Ambient* **21**(1): 68–73.

Cummins H Z and Swinney H L (1970) Light beating spectroscopy. *Prog. Opt.* **8**: 133–200.

DaSilva E R and Muramatsu M (2008) Comparative study of analysis methods in biospeckle phenomenon. *AIP Conf. Proc.* **992**: 320–325.

Enes A M, Fracarolli J A, Dal Fabbro I M and Rodrigues S (2012) Biospeckle supported fruit bruise detection. *Int. J. Biol. Biomol. Agric. Food Biotechnol. Eng.* **6**: 10–24.

Fujii H, Asakura T, Nohira K, Shintomi Y and Ohura T (1985) Blood flow observed by time-varying laser speckle. *Opt. Lett.* **10**: 104–106.

Fujii H, Nohira K, Yamamoto Y, Ikawa H and Ohura T (1987) Evaluation of blood flow by laser speckle image sensing. *Appl. Opt.* **26**: 5321–5325.

Gao Y and Rao X (2019) Black spot bruise in potatoes: Susceptibility and bio speckle activity response analysis. *J. Food Meas. Charact.* **13**: 444–453.

González-Peña R J, Braga R A, Cibrián R M, Salvador-Palmer R, Gil-Benso R and San Miguel T (2014) Monitoring of the action of drugs in melanoma cells by dy- namiclaser speckle. *J. Biomed. Opt.* **19**(5): 1–5.

Isis C A, Braga R A, Ramos E M, Ramos A L S and Roxael E A R (2013) Application of bio speckle laser technique for determining biological phenomena related to beef aging. *J. Food Eng.* **119**: 135–139.

Jitendra D, Litesh B, Vimal B and Shashi P (2017) Advances in optical science and engineering. *Springer Proc. Phy.* **194**: 389–394.

Kumari S and Nirala A K (2016) Bio speckle technique for the non-destructive differentiation of bruised and fresh regions of an Indian apple using intensity-based algorithms. *Laser Phy.* **26**: 115–161.

Kurenda A, Adamiak A and Zdunek A (2012) Temperature effect on apple bio speckle activity evaluated with different indices. *Postharvest Biol. Technol.* **67**: 118–123.

Lammertyn J, Peirs A, De Baerdemaeker J and Nicolas B (2000) Light penetration properties of NIR radiation in fruit with respect to non-destructive quality assess ment. *Postharvest Biol. Technol.* **18**(2): 121–132.

Lasio B and Zude M (2009) Analysis of laser light propagation in kiwi fruit using backscattering imaging and Monte Carlo simulation. *Comput. Electron. Agric.* **69**(1): 33–39.

Lee S, Lonumi S, Lim H S, Goron I, Cno B K, Kim M S, et al (2015) Development of a detection method for adulterated onion powder using Raman spectroscopy. *J. Fac. Agric.* **60**(1): 151–156.

Mandracchia B, Palpacuer J, Nazzaro F, Bianco V, Rega R, Ferraro P, et al (2018) Biospeckle de correlation quantifies the performance of alginate-encapsulated pro- biotic bacteria. *IEEE J. Sel. Top. Quantum Electron.* **25**(1): 1–6.

Matheus S, Silva K G, Fujii A K, Poppi R J and Fracarolli J A (2017) Sugarcane (Saccharum officinarum L.) analysis through biospeckle and spectroscopy (NIR). *J. Agric. Sci. Technol.* **7**: 62–68.

Mesa K, Serra S, Masia A, Gagliardi F, Bucci D and Musacchi S (2016) Seasonal trends of starch and soluble carbohydrates in fruits and leaves of 'Abbé Fétel'peartrees and their relationship to fruit quality parameters. *Sci. Horticul.* **211**: 60–69.

Minz P D and Nirala A K (2016) Laser speckle technique to study the effect of chemical pre-treatment on the quality of minimally processed apples. *Laser Phy.* **26**: 1–8.

Mulone C, Budini N, Vincitorio F M, Frevre C E, Lopez A J and Ramim A (2014) Biospeckle activity evolution of strawberries. *SOP Trans. Appl. Phys.* **1**(2): 65–73.

Murialdo S, Sendra G H, Passoni L, Arizaga R, Gonzalez J F, Rabal H J, et al (2009) Analysis of bacterial chemotactic response using dynamic laser speckle. *J. Biomed. Opt.* **14**: 254–258.

Murray K, Wu F, Shi J, JunXue S and Warriner K (2017) Challenges in the mi crobiological food safety of fresh produce: Limitations of post-harvest washing and the need for alternative interventions. *Food Qual. Saf.* **1**(4): 289–301.

Oleksandr P, Muravsky L I and Berezyuk M I (2015) Application of bio-speckles for assessment of structural and cellular changes in muscle tissue. *J. Biomed. Opt.* **20**(9): 0950061–0950067.

Pajuelo M, Baldwin G, Rabal H, Cap N, Arizaga R and Trivi M (2003) Bio speckle assessment of bruising in fruits. *Opt. Lasers Eng.* **40**: 13–24.

Pengcheng L, Ni S, Zhang L, Zeng S and Luo Q (2006) Imaging cerebral blood flow through the intact rat skull with temporal laser speckle imaging. *Opt. Lett.* **31**: 1824–1826.

Pieczywek P M, Cybulska J, Szymańska-Chargot M, Siedliska A, Zdunek A, Nosalewicz A and Kurenda A (2017) Early detection of fungal infection of stored apple fruit with optical sensors Comparison of bio speckle, hyper spectral imaging and chlorophyll fluorescence. *Food Control.* **85**: 327–338. https://doi.org/10.1016/j.foodcont.2017.10.013.

Piotr M P, Nowacka M, Dadan M, Wiktor A, Rybak K and Witrowa-Rajchert D (2018) Postharvest monitoring of tomato ripening using the dynamic laser speckle. *Sensors.* **18**: 1093.

Rabal H J (2008) Dynamic laser speckle and applications. *Optik.* **128**: 501–507.

Rabelo G F, Enes A M, Junior R A B and DalFabbro I M (2011) Frequency response of biospeckle laser images of bean seeds contaminated by fungi. *Biosyst. Eng.* **110**(3): 297–301.

Ramos B, Miller F A, Brandão T R, Teixeira P and Silva C L (2013) Fresh fruits and vegetables—an overview on applied methodologies to improve its quality and safety. *Innov. Food Sci. Emerg. Technol.* **20**: 1–15.

Retheesh R, Ansari M Z, Radhakrishnan P and Mujeeb A (2018) Application of qualitative bio speckle methods for the identification of scar region in a green orange. *Mod. Phys. Lett.* **32**: 1850113.

Samuel B, Retheesh R, Ansari M Z, Nampoori V P N, Radhakrishnan P and Mujeeb A (2017) Cross-correlation and time history analysis of laser dynamic speckle gram imaging for quality evaluation and assessment of certain seasonal fruits and vege tables. *Laser Phys.* **27**(10): 105601–105607.

Silva S H, Lago A M T, Rivera F P, Prado M E T, Braga R A and deResende J V (2018) Measurement of water activities of foods at diff erent temperatures using biospeckle laser. *J. Food Meas. Charac.* **12**(3): 2230–2239.

Sutton D B and Punja Z K (2017) Investigating biospeckle laser analysis as a diagnostic method to assess sprouting damage in wheat seeds. *Comput. Electron. Agric.* **141**: 238–247.

USEPA (2015) EDSP weight of evidence conclusions on the tier 1 screening assays for the list chemicals US environmental protection agency report. https://www.epa.gov/sites/production/files/2015-08/documents/edsp_comprehesive_management_plan_021414_f.pdf (Accessed 8 June 2019).

Vladimirov A P, Novoselova I A, Mikhailova Y A, Bakharev A and Yakin D I (2016) The use of laser dynamical speckle interferometry in the study of cellular processes. *J. Biomed. Photonics Eng.* **2**(1): 1–6.

Chapter 7

Cold Plasma for Decontamination of Food

Saadiya Naqash, H. R. Naik, Darakshan Majid, H. A. Makroo, and B. N. Dar

7.1 Introduction

In the current era of consumer awareness of the security and nutritional viability of different food products, food researchers are being diverted towards novel processing techniques that can help provide safe and nutritionally sustainable food products without comprising the flavour, texture and appearance of processed foods. Originally, food processing involved the use of heat treatments (pasteurization, sterilization, blanching etc.) for decontamination. Although these classic methods of food decontamination reach a satisfying level of safety, they involve a longer processing period and high input of energy, thereby tending to be economically prohibitive. These treatments also influence the overall features of foodstuffs by decreasing the bioactivity of heat-labile nutrients (Jouany, 2007). Owing to the flexible and diverse nature of food industries to incorporate new and profitable technologies, the attention of research bodies and regulatory agencies has been attracted by novel food processing techniques termed "non-thermal food processing techniques" that generally involve lower heat for food preservation in addition to minimizing nutrient losses. These non-thermal

DOI: 10.1201/9781003217138-7

techniques include pulse electric field technology (Zhao et al., 2012), high hydrostatic pressure processing (Eisenmenger and Reyes-De-Corcuera, 2009), ultrasonic treatment (O'Donnell et al., 2010), ultraviolet light decontamination (Koutchma, 2009) and gamma irradiation (Kuan et al., 2013) and have been widely used for various enzyme inactivation as well as microbial decontamination of food products.

Considering the novelty and non-thermal nature of these techniques, researchers across the globe have been fascinated regarding their vast potential usage. Besides, they are efficient, environmentally friendly and non-invasive in nature. New research suggests cold plasma is a highly influential and cost-effective technology in various food processing techniques (Charoux et al., 2021). Previously, this technique was used to restructure various functional properties of polymers and increase the surface energy of several products besides having a huge number of uses in electronics. In addition, it is widely used in textiles, packaging (glass, paper) and various other industries (Ekezie et al., 2017).

In 1923, Irving Langmuir coined term "plasma", which is generally characterized as a fourth addition to the existing three states of matter. It contains partial or entirely ionized gases comprising numerous reactive species like free radicals; reactive species; and elements of nitrogen, oxygen and hydrogen that are either in equilibrium or non-equilibrium (Varilla et al., 2020). Plasma is generally categorized into thermal and non-thermal (Figure 7.1) depending on its thermodynamic state of equilibrium (Zhang et al., 2017). In thermal plasma, a thermal equilibrium exists between the reactive species (charged and neutral) (Pankaj et al., 2013).

Cold plasma technology involves gas in an energized state and low temperature and atmospheric pressure, making it a potential technique for use in food processing and development. The energized gas produced by applying electric force at low temperature and pressure to the gas generates plasma with a huge quantity of free radicals that encounter the microorganisms and biomolecules (enzymes) in food products, causing their deactivation by changing their structural orientation and functionality. Due to the limited life span of these reactive particles, no residue is left at the end of

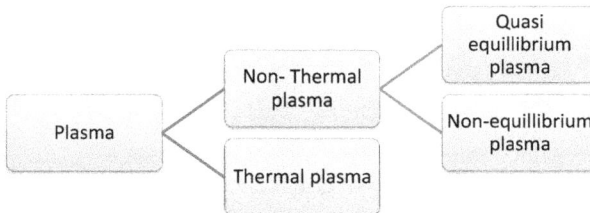

Figure 7.1 Different types of plasma based on thermal equilibrium.

the process; therefore, it is considered green technology. Also, cold plasma involves less treatment time, thereby reducing the energy utilization and negligible quality changes without any compromise in food security and safety (Bourke et al., 2018).

In recent times, cold plasma technology is widely employed in food industries for various purposes like enhancement in seed germination and growth of plants (Randeniya and de Groot, 2015), water and soil treatment/decontamination (Sarangapani et al., 2017; Zhang et al., 2017), cleansing of foods and food processing equipment (Misra et al., 2015; Leipold et al., 2010), inactivation of biomaterials and enzymes (Misra et al., 2016) and modification of food packaging properties (Pankaj et al., 2014). However, the major challenge of using the cold plasma technique is the source of plasma generation for which a complete understanding of the use of optimized techniques for the different process is required. Considering the novelty, potential and non-invasive nature of this technology, this chapter illustrates the fundamentals, chemistry, decontamination capacity and sustainability of this process besides addressing its limitations and future scope.

7.2 Fundamentals of Cold Plasma Technique

The three existent states of matter, solid, liquid and gas, are familiar terms to every person. These three states are interchangeable by employing different processes, as shown in Figure 7.2. Solid (ice) changes to liquid while

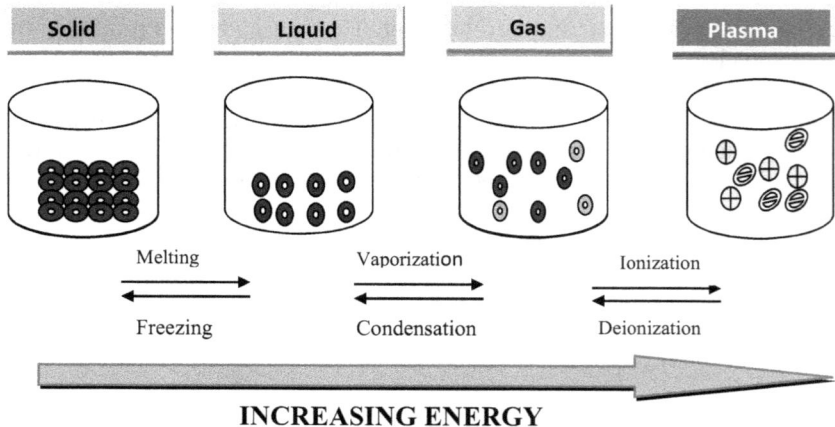

Figure 7.2 Transition of different states of matter with respect to increasing energy.

heat is applied to it; furthermore, this liquid can be changed to a gaseous state on application of further temperature and pressure. However, a question is: what if the pressure and temperature are further increased to the gas? When the gas is energized to a larger extent, there occurs a severe molecular breakdown, resulting in the generation of several reactive species (electrons, ions, free radicals, etc.). At this stage, this energized gas is termed plasma and consists of ions (positive and negative) mostly in the same quantity, making it a neutral form, as well as containing free radicals, electrons, photons and other molecules (Coutinho et al., 2018).

Besides using thermal energy, the generation of ionized gas is also possible by employing high electric fields. A prominent electric field is used for the ionization of a large volume of gas, resulting in the formation of charged particles. The electrons generated gain maximum energy because of their lower weight, thereby achieving a very high temperature of approximately 10^3–10^4 Kelvin. The plasma generated at this temperature is termed thermal plasma. However, the plasma generated at ambient temperature under atmospheric pressure is termed non-thermal plasma, which is generated using electric discharge only, and the temperature of this plasma is close to ambient temperature, under 60°C. (Misra et al., 2019). The heavy species in cold plasma have a lower temperature than that of the electrons, resulting in inconsistent moment transfer when they collide with each other under the effect of the difference in electric potential (Misra et al., 2018). In the modern era, there is a major advancement in plasma technology that involves the use of high-voltage or high-frequency power systems. The plasma generated at this stage is termed atmospheric pressure plasma and has potential applications in food processing industries.

7.3 Sources of Cold Plasma

There are several sources of plasma, and the main aim of all the sources is to take advantage of the potential and advanced techniques of gas ionization for the planned usage. Consequently, several types of plasma sources are used. The equipment configuration used to provide an electric field for the generation of plasma is termed a plasma source (Misra and Roopesh, 2019). Different kinds of plasma sources have been developed by researchers. A brief summary of a few that are used for decontamination of food and bio-products is given in subsequent sections.

7.3.1 Corona Discharge

Corona discharge plasma is generated using a strong electric field that surrounds a conducive atmosphere during the application of high voltage. The gas is ionized due to electrical breakdown caused by a high electric field that surrounds the needle (Figure 7.3), generating a self-sustaining, radiant fountain-like discharge and helping in uniform stabilization of discharge over the surface of electrodes (Misra and Roopesh, 2019).

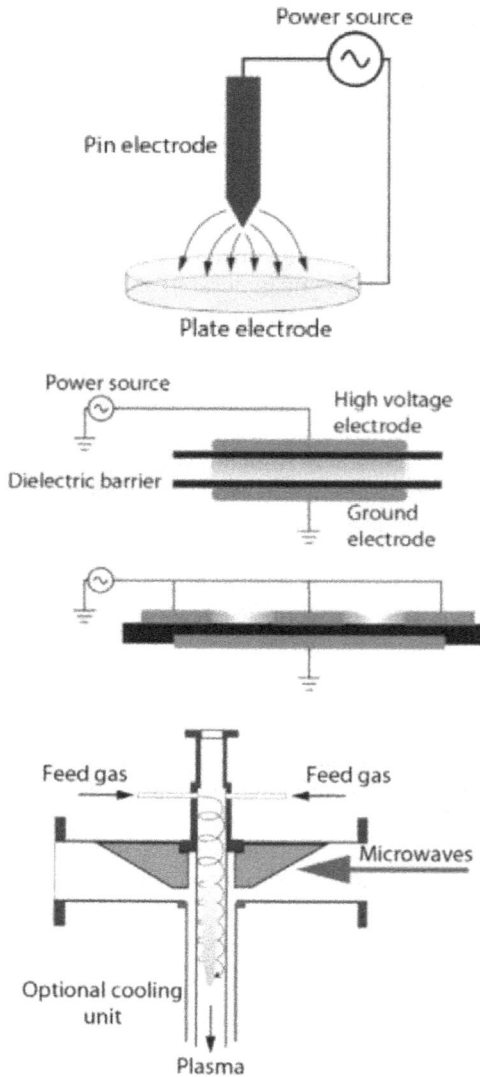

Figure 7.3 Different sources of plasma generation.

Source: Adapted from Misra and Roopesh (2019)

7.3.2 Dielectric Barrier Discharge

The setup of dielectric barrier discharge consists of two electrodes separated from each other by a dielectric material at a distance of 10^{-3} to 10^{-2} m, and the space between the two electrodes is surrounded by the gas. Of the two electrodes, one is exposed to a high electric field (10^2 to 10^4 V), and the second is grounded or sometimes left floating. Due to the increasing electric voltage, the gas between the spaces is ionized, resulting in the generation of plasma. Depending on the application of voltage, the dielectric barrier discharge (DBD) could shift to an arc regime, glow regime or microfilamentary discharge. Plasma technology used for the decontamination of packaged food materials employed DBD plasma technology, wherein the packaging material acts as a dielectric barrier (Misra et al., 2019). Besides the parallel plate configuration, the DBD setup could be prepared using two concentric cylindrical electrodes. These electrodes form a source where a plasma jet is generated.

7.3.3 Microwave Plasma

In addition to corona discharge and DBD plasma sources, there is another source of plasma generation that employs microwave energy (915 to 2450 MHz) for gas ionization (Leins et al., 2014). Figure 7.3(c) represents a microwave plasma setup where, due to varying movements of electrons and ions, an alternating current is generated that increases the power of electrons, thereby resulting in the generation of cold plasma at high gas pressures. A detailed description of different setups of plasma sources is complex and is beyond the scope of this chapter. Commonly, the selection of different plasma sources depends on the type of application for accomplishing the specific goal.

7.4 Chemistry of Plasma Generation

Plasma chemistry is an intricate and complex subject that involves different reaction mechanisms resulting in the generation of active plasma that consists of many reactive species (Van Gaens and Bogaerts, 2013). Plasma chemistry depends on various factors like the composition of the gas, electric field power and frequency. Different reactive species that are formed and are majorly related to the antimicrobial properties of plasma, including hydrogen peroxide, hydronium ions, ozone and hydroxyl radicals, nitrogen oxide superoxide anions and hydroperoxyl (Arjunan et al., 2015). There are several other compounds, free ions and vibrationally and energetically excited molecules, that contribute to the

Table 7.1 Comprehensive Listing of Fundamental Free Radicals and Ground State Species Relevant to Antimicrobial Activity in Non-Equilibrium Non-Thermal Humid Air Plasma (Adapted from Herron and Green, 2001)

Oxygen	Hydrogen	Nitrogen	HO_x	No_x	HN_x
$O+O_3 \rightarrow O_2+O_2$	$H+HN \rightarrow N+H_2$	$N+O_2 \rightarrow NO+O$	$HO+H_2 \rightarrow H_2O+H$	$NO+O_3 \rightarrow NO_2+O_2$	$HN+NO \rightarrow$
$O+NO_2 \rightarrow O_2+NO$	$H+O_3 HO \rightarrow +O_2$	$N+O_3 \rightarrow NO+O_2$	$HO+O_3 \rightarrow HO_2+O_2$	$NO+NO_3 \rightarrow 2NO_2$	$H+N_2O$
$O+NO_3 \rightarrow O_2+NO_2$	$H+HO_2 \rightarrow$	$N+NO \rightarrow N_2+O$	$HO+HO \rightarrow H_2O+O$	$NO+HO_2 \rightarrow$	$HN+NO_2 \rightarrow$
$O+N_2O_5 \rightarrow$ products	$HO+HO$	$N+NO_2 \rightarrow$	$HO+HO_2 \rightarrow H_2O+O_2$	$HO+NO_2$	products
$O+HN \rightarrow H+NO$	$H+HNO \rightarrow H_2+NO$	N_2O+O	$HO+H_2O_2 \rightarrow HO_2+H_2O$	$NO+HNO$	$HN+O_2 \rightarrow$
$O+HO \rightarrow O_2+H$	$H+HONO \rightarrow$	$N+NO_3 \rightarrow$	$HO+HNO \rightarrow NO+H_2O$	$HO+N_2O$	$HO+NO$
$O+HO_2 \rightarrow O_2+HO$	H_2+NO_2	$NO+NO_2$	$HO+HONO \rightarrow$	$NO_2+O_3 \rightarrow$	$HN+HO \rightarrow$
$O+H_2O_2 \rightarrow HO_2+$	$H+NO_2 \rightarrow$	$N+HN \rightarrow H+N_2$	NO_2+H_2O	NO_3+O_2	$NO+H_2$
HO	$HO+NO$	$N+HO \rightarrow NO+H$	$HO+HONO_2 \rightarrow$		$HN+HO_2 \rightarrow$
$O+HNO \rightarrow NO+HO$		$N+HO_2 \rightarrow NO+$	NO_3+H_2O		products
$O+HONO \rightarrow NO_2+$		HO	$HO+NO_3 \rightarrow HO_2+NO_2$		Major atomic
HO			$HO+N_2O_5 \rightarrow$ products		oxygen reactions
$O+ \rightarrow HONO_2$			$HO+HO_2NO_2 \rightarrow$ products		$H_2+O \rightarrow H+OH$
Products			$HO_2+HO_2 \rightarrow H_2O_2+H_2O$		$CH_4+O \rightarrow CH_3+$
$O+ \rightarrow HO_2NO_2$			$HO_2+O_3 \rightarrow HO+2O_2$		OH
products			$HO_2+NO_3O_2+ \rightarrow$		H_2O+O $2OH$
			NO_2+HO		O_2+ $OO_3 \rightarrow$
			$HNO+NO_2 \rightarrow$		$CO+O \rightarrow CO_2$
			$HONO+NO$		
			$NO_2+NO_3 \rightarrow$		
			$NO+NO_2+O_2$		
			$NO_3+NO_3 \rightarrow 2NO_2+O_2$		
			$NO_3+O_2 \rightarrow$ products		
			$N_2O_5+H_2O \rightarrow 2HONO_2$		

antimicrobial activity of plasma. A detailed study of plasma chemistry can be found in a review by Lu et al. (2016).

Table 7.1 shows different reaction mechanisms that take place during the antimicrobial activity of the plasma technique. The category, quantity and strength of the reactive species depend on the type and frequency of electrical power used in addition to the source employed for plasma generation.

7.5 Application of Cold Plasma for Food Decontamination

In recent years, cold plasma technology has come up as a potential food processing application that caters to a large number of processing issues. Cold plasma technology employed in the food industry has wide applications, such as food decontamination, processing and functional modification. Figure 7.4 is a graphical illustration of cold plasma technology application in food industries.

Food safety is a major challenge because of the occurrence of various food contaminants like mycotoxins (Mousavi Khaneghah et al., 2018), pesticides (Razzaghi et al., 2018), allergens (Schaarschmidt, 2016; Sicherer and

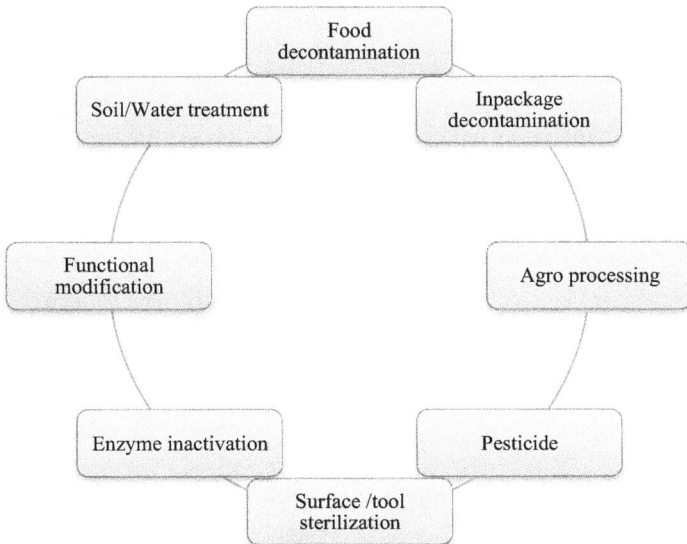

Figure 7.4 Application of cold plasma in food industries.

Sampson, 2018) and polycyclic aromatic hydrocarbons (PAHs) (Yousefi et al., 2018). The risks associated with foodborne infections being transmitted are a universal issue, and various methods are employed to take care of them.

Over the past few decades, cold plasma technology has gained prominence as a potential decontamination technique (Gavahian et al., 2020; Gavahian et al., 2019). Extensive research on the antimicrobial properties of plasma technology has established it as a prominent source of pathogenic microbes, fungus, mould, mycotoxins and enzyme inactivation and degradation (Gavahian and Cullen, 2020). In addition, numerous studies have been conducted regarding the degradation of pesticides due to generation of highly reactive species that degrade the chemical structure of pesticide molecules, thereby rendering them ineffective (Gavahian and Khaneghah, 2020). Many researchers have established successful usage of cold plasma technology for disinfection of many food products. In a survey conducted by Pignata et al. (2017), it was reported that over the past decade, 40% of plasma treatment is done on fresh fruits and vegetables for disinfection purposes, followed by 21% on nuts, seeds and other dried fruit. Meat and meat products are also plasma treated at a level of 19%, followed by 10% on spices, 6% on liquids and beverages and 4% on eggshells.

7.5.1 Antimicrobial Activity

Antimicrobial activity of cold plasma is attributed to the detrimental effect of scavenging free radicals that are generated in the plasma. These reactive species cause functional damage to various biomolecules like proteins, enzymes and even genetic material (DNA and RNA). Figure 7.5 shows cellular damage induced by cold plasma.

Various studies have suggested that when bacterial cells are exposed to an intense electric field, there is a rupture of bacterial cell membranes (Mendis et al., 2000). Additionally, cold plasma exposure leads to complete destruction of the outer bacterium membrane (Yusupov et al., 2012). Han et al. (2016) reported the degradation mechanism of Gram-positive and Gram-negative bacteria by using the cold plasma technique. They established that cold plasma–induced inactivation of Gram-positive bacteria like *S. aureus* results due to internal cell damage, while in Gram-negative bacteria like *E. coli*, damage is attributed to the leakage of cell constituents and damage of DNA.

7.5.2 Fungal and Mycotoxin Decontamination

Fungi belong to a class of uninucleated single-cell or multicellular organisms that are larger than bacteria and mostly infest the food chain via soil, seeds, water or airborne pathogenic spores.

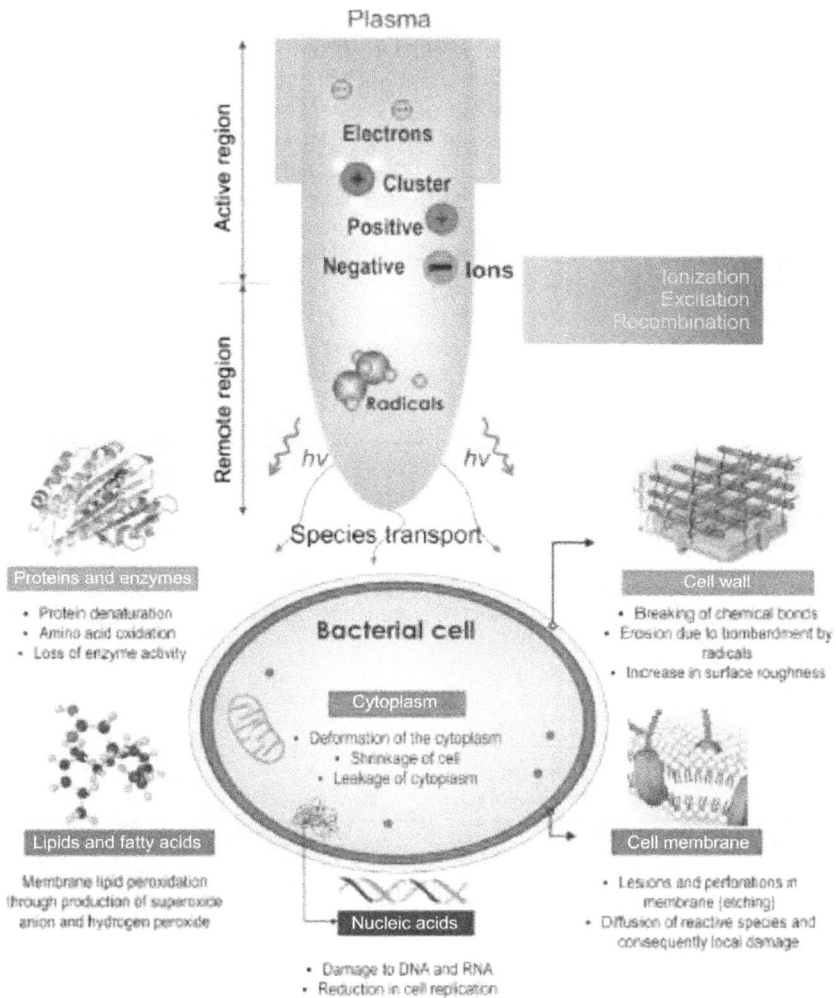

Figure 7.5 Cold plasma–induced microbial cell damage.

Source: Adapted from Misra and Jo (2017)

Mycotoxins are fungal secondary metabolites that pose toxicity to both humans and animals. These metabolites can be carcinogenic, genatotoxic and mutagenic. In order to alleviate concerns about them, several researchers have done exhaustive studies regarding the degradation of mycotoxins using cold plasma, and significant degradation has been shown in various food products (Ten Bosch et al., 2017). As documented by Devi et al. (2017), exposure of infested nuts to plasma for 30 minutes at 60 W reduces the spore content of *Aspergillus flavus* and *Aspergillus parasiticus* by 99.9% and 99.5%, respectively. Some authors also reported that plasma treatment results in a 96.8% reduction in aflatoxin B1 (AFB_1), which is a highly toxic mycotoxin and the most powerful naturally occurring carcinogen produced from *A. flavus*, and up to approximately 95% reduction in AFB_1 production from *A. parasiticus*. In a study done by Dasan et al. (2016), an atmospheric plasma fluidized bed reactor is used for sanitization of maize against *Aspergillus flavus* and *Aspergillus parasiticus* that resulted in more than a 5 log reduction. Hojnik et al. (2017) applied argon plasma on date palms that resulted in the inhibition of mycotoxins like fumonisin B2 and ochratoxin A. Table 7.2 shows few recent research results on disinfection of food products using cold plasma technology.

The method of decrease in mycotoxin in food products can be explained either as interruption of fungal activities or complete eradication of toxins. Exposure of food products to plasma leads to a reduction in mycotoxins and cessation of proliferation by mycotoxin-producing fungi. As reported by Dasan et al. (2016), it was established that exposure of hazelnuts to atmospheric cold plasma at a frequency of 25 kHz and 700 W power is optimum for spore reduction of *A. parasiticus* and *A. flavus* by log 4.5 and

Table 7.2 Decontamination of Food Products Using Cold Plasma Technology Based on Recent Studies

Microbe	Processing Parameters	Key Findings	Reference
E. coli O157:H	Low-pressure cold plasma treatment at 50 W for 20 minutes	Significant ($p > 0.05$) reduction in microbial load of dried pepper mint (*Mentha piperita L.*)	Kashfi et al., 2020
Alternaria mycotoxin	Dielectric barrier discharge for 3–5 min at 30 to 40 kV	Both alternariol and alternariol monomethyl ether in aqueous solution as well as solid state could effectively be removed by DBD cold plasma	Wang et al., 2020

Microbe	Processing Parameters	Key Findings	Reference
Campylobacter and Salmonella	Dielectric barrier discharge for in-package cold plasma treatment at 35% oxygen and 65% carbon dioxide at 60 kV for 60 s	Reduction of microbial load by 2 log	Zhuang et al., 2020
Aspergillus flavus	Atmospheric cold plasma at 9 kV for 6 min	Reduction in Aspergillus flavus by 2.98 log CFU/g	Bagheri et al., 2020
AFB1, OTA, ENB, FB1, ZEN, DON	Source (plasma jet, 6 kV), frequency (20 kHz), time (up to 60 min), plasma gases (helium with 0.5% and 0.8% oxygen gas)	Reduction of all investigated mycotoxins	Wielogorska et al. (2019)
E. coli	Corona discharge at 20 kV 1 atm, frequency, 58 kHz for 2 min	Reduction in microbial load on surface of pork loin by 1 log CFU/gm	Choi et al., 2018
Listeria monocytogenes	Dielectric barrier discharge at 100 W, frequency of 15 kHz for 10 min	Reduction in microbial load in pork butt by 3.86 log CFU/gm	Jayasena et al. (2015)

(Continued)

Table 7.2   (Continued)

Microbe	Processing Parameters	Key Findings	Reference
Staphylococcus aureus	Dielectric barrier discharge at 1 atm pressure and radio frequency at 30 kV power for 10 min	Reduction in microbial load of beef jerky by 4.00 log CFU/g	Kim et al. (2015)
Aerobic mesophilic microbes	Dielectric barrier discharge at frequency of 50 Hz and 60 kV for 5 minutes	Reduction of microbial load by 2 log	Misra et al. (2014)

log 4.2, respectively. Degradation of aflatoxin by cold plasma was reported by Shi et al. (2017), where they established that higher relative humidity, longer exposure times and optimum gas mixture (65%, 30% and 5% of oxygen, carbon dioxide and nitrogen, respectively) can increase the aflatoxin degradation rate. Exposure of peanuts to cold plasma has been reported to eradicate aflatoxins like B1, B2, G1 and G2 (Devi et al., 2017). They also explained that increased exposure times and higher powers increase the decontamination properties of plasma process significantly; however, the decontamination process majorly depends on the type of mycotoxin. Similar studies were conducted by Šimončicová et al. (2018) confirming the massive inactivation of *A. flavus hyphae* in groundnuts when they were exposed to a cold plasma treatment for 24 minutes at an electric power of 60 W. As reported by Gavahian and Cullen (2020), the degradation rate of mycotoxins differs according to their structure and extracellular matrix, thereby converting them into harmless secondary components. Growth inhibition of *A. flavus* on cereal bars, brown rice and malt extract agar media using jet plasma operating at a power of 40 W for 25 and 20 min, respectively, was reported by Suhem et al. (2013). Plasma-induced degradation of yeast and mould in blueberries was studied by Lacombe et al. (2015). They reported that exposing inoculated blueberries to plasma treatment for a duration of 15 to 120 s resulted in a reduction of yeast and mould by 0.8 to 1.6 log CFU/g. Misra et al. (2014) also reported a reduction by 3.3 log cycle of naturally occurring yeast and moulds on strawberries when exposed to cold plasma

treatment for 5 min at 60 kV power. The ability of cold plasma exposure to decrease the population of yeast and mould on pepper seeds and oregano was investigated by Hertwig et al. (2015), who reported that exposure of these products to cold plasma treatment for 60 min in crushed oregano and for 5 min in black pepper seeds reduced mould by 1.8 log 10 CFU/g. In a similar study by Kim et al. (2017), the changes caused due to exposure time, electrical power and type of gas were investigated for decontamination of red pepper by employing a microwave-assisted plasma system, and a reduction of *A. flavus* by 2.5 log was observed by using nitrogen gas at a power of 900 W for 20 min.

In order to bring down the yeast and mould population in meat and meat products, Ulbin-Figlewicz et al. (2015) investigated the effect of helium-, argon- and nitrogen-generated plasma exposure for 10 min on growth of yeast and mould in meat and reported a reduction of 3, 2.6 and 1 log CFU/cm^2, respectively. In another study by Yong et al. (2017), inactivation of *A. flavus* on beef was reported when the product was exposed to plasma treatment for 10 minutes, reducing the mould load to 2.06 log CFU/g from an initial load of 5.24 log CFU/g. In most countries, mould contamination of dried fish products is one of the major challenges that have been addressed by cold plasma treatment for 20 minutes, thereby reducing the mould count by 1 to 1.5 log CFU/g (Park and Ha, 2015).

According to a study done by Pankaj et al. (2018), the mycotoxin degradation mechanism of plasma treatment is mainly related to the molecular structure of mycotoxins, type of plasma, reactive species and free radicals that include ozone, hydroxyl radicals, radicals from ionized gas and the CO_2 precursors produced that come into contact with the toxin. Figure 7.6 shows the suggested pathway of mycotoxin degradation by plasma, which involves several addition and degradation reactions, resulting in the inactivation of moulds.

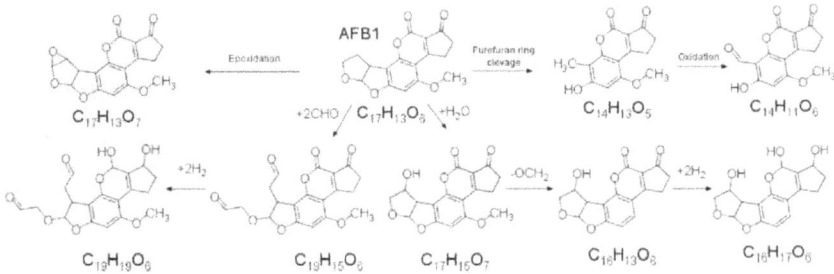

Figure 7.6 Aflatoxin B1 degradation pathway by atmospheric cold plasma treatment.

Source: Adapted from Shi et al. (2017)

7.5.3 Bacterial Decontamination

The production and nature of food and its products make it vulnerable to bacterial growth and development. Significant and persistent efforts are being carried out for combating these challenges, paving the way for new approaches and technologies. As discussed in the preceding sections, cold plasma helps in decontamination of food products against all pathogenic microbes.

Bacterial contamination is a phenomenon that can take place at any stage from farm to fork and is one of the major reasons for serious disease outbreaks in humans. The action mechanism of cold plasma depends on ionized gas and the presence and intensity of reactive oxygen species and free radicals that attack microbial cells. In general, bacteria, particularly Gram-positive bacteria, are highly vulnerable to the attack of reactive species (Stoffels et al., 2008), as they cause cytoplasmic membrane and genetic damage to the bacterial cells. As reported by Joshi et al. (2011), free radicals cause lipid peroxidation in bacterial cell membranes. On severe exposure to charged species, the chemical bonds in bacterial cells are broken, resulting in attrition, lesion formation and leakage; however, in Gram-negative bacteria, damage occurs easily, unlike in Gram-positive bacteria, due to the presence of a thicker structural membrane (Moreau et al., 2008). Perforations in cell membranes caused by etching enhance the diffusion of other reactive species formed during plasma discharge and cause bond breakages that result in morphological alteration in bacterial cells, causing their complete inactivation (Fricke et al., 2012). In an investigation done by Hertwig et al. (2017), decontamination of unpeeled almonds against *Salmonella enteritis* it was reported that using different gases (air, oxygen, nitrogen, carbon dioxide and combination of air and carbon dioxide) displayed different rate of bacterial inactivation with a maximum reduction of 5 log CFU/g achieved by air plasma when the samples were exposed for 15 min.

In a recent work by Zhuang et al. (2020), optimization of cold plasma technology revealed that treatment of chicken breast inoculated with *Campylobacter* and *Salmonella* with in-package cold plasma (IPCP) resulted in inactivation of psychrophiles by more than 1 log CFU/g when the conditions were maintained with 35% oxygen and 60% carbon dioxide at 60 kV for 60 s. in a similar study by Moutiq et al. (2020), it was suggested that IPCP treatment at a voltage as high as 100 kV results in much higher reduction (> 2 log CFU/g) of natural microflora in lean poultry meat surfaces without altering the overall quality.

Fruits and vegetables are an important part of the human diet and are highly vulnerable to bacterial and microbial infestation. In a study by Berardinelli et al. (2016), treatment of pear with cold plasma resulted in a significant reduction of mesophilic bacteria, amounting to up to 2–5 log CFU per fruit. Similarly, exposure of dragon fruit to cold plasma treatment at 40 W resulted in the inactivation of *E. coli* and *S. typhimurium*, as reported by Matan et al. (2015). Misra et al. (2016) reported that damage to genetic material on exposure to cold plasma is due to several mechanisms, including thymine dimer formation, alteration in nucleobases and oxidation of nucleotides by reactive species. Degradation of DNA has been found to be a major factor that results in the antibacterial effect of cold plasma of product surfaces infested with microbes like *Bacillus cereus*, *Staphylococcus aureus*, *E. coli* and *Pseudomonas* sp. In a study conducted by Kashfi et al. (2020), exposure of dried peppermint to low-pressure cold plasma at 50 and 60 W for 20 minutes resulted in elimination of *E. coli* and *Salmonella* species. In conclusion, Mai-Prochnow et al. (2014) suggested that the antibacterial effect of cold plasma is due to permeabilization of the bacterial cell walls that results in complete damage of intracellular organelles and the protein matrix.

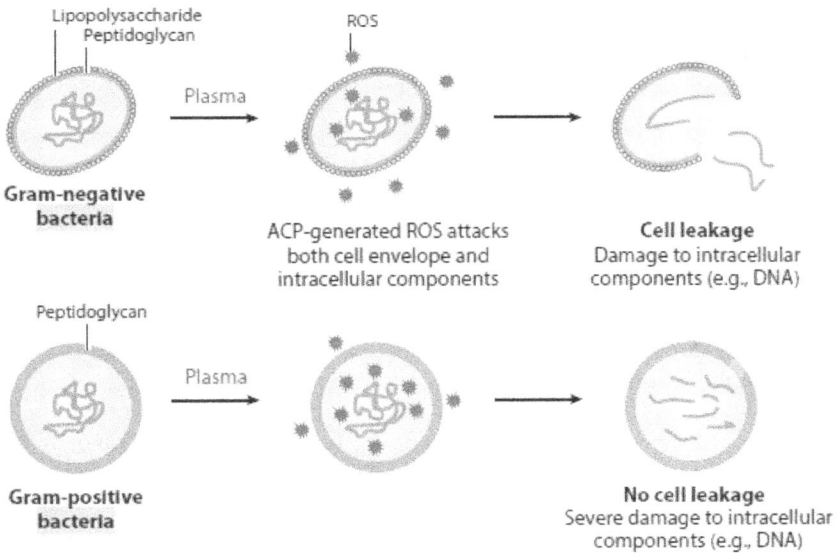

Figure 7.7 Mode of action of ACP on Gram-positive and Gram-negative bacteria.

Source: Adapted from Sarangapani et al. (2018)

7.5.4 Viral Decontamination

Viral disinfection is one of the foremost challenges in the food supply chain, and in the modern era, cold plasma has emerged as a promising virucidal agent. The antiviral capacity of the cold plasma technique has been extensively studied by many researchers, who report that cold plasma has the capacity to inactivate bacteriophages swiftly. In addition, viral reduction approximating 4–6 log CFU/ml has been reported by exposure to cold plasma for a period of 10 minutes (Alshraiedeh et al., 2013; Venezia et al., 2008). Lambda phage is a facile viral model that infects humans and plants alike, and its degradation or inactivation by exposure to cold plasma for 20 s was reported by Nakashima et al. (2010). A few other studies have reported the virucidal capacity of different sources of plasma against different viruses. such as non-thermal plasma jet against the MS2 bacteriophage, resulting in its inactivation (Alkawareek et al., 2012). Inactivation of nebulized respiratory human parainfluenza virus, (RSV) and influenza virus has been well reported in various experimental studies. Norovirus, a toxic food poisoning outbreak strain, has reportedly been inactivated by cold plasma, thereby enhancing the potential of the cold plasma technique to be used as an important virucidal agent to prevent the risk of further transmission of infection (Ahlfeld et al., 2015). These studies reveal that cold plasma could establish promising virucidal activity and anti-infectious environment over a wide range of pathogenic diseases.

7.5.5 Pesticide Decontamination

According to statistics from the European Union's Pesticide Action Network, the presence of more than 200 different pesticides in food in Europe has been reported and is a major global concern. The toxicity of pesticide residues is responsible for various serious diseases like cancer, asthma and cardiovascular complications (Figure 7.7). Therefore, it is essential to employ various effective and sustainable ways that can eliminate or significantly reduce pesticide residues in foods (Sarangapani et al., 2018). Degradation of pesticides by cold plasma treatment is ascribed to a multitude of reactive species that can facilitate the oxidation of pesticide residues. Reduction of pesticide remains in water and on the surface of fresh strawberries using cold plasma treatment for 5–8 minutes was reported by Sarangapani et al. (2016) and Misra et al. (2014), establishing that degradation of pesticide residues into much simpler, harmless compounds, thereby offering the potential for zero-residue clean label processing. Pesticides like dichlorvos and omethoate on maize plants are reported to be degraded by using oxygen plasma, establishing that the effectiveness of pesticide degradation depends majorly on treatment parameters and the functional and chemical structure of pesticides.

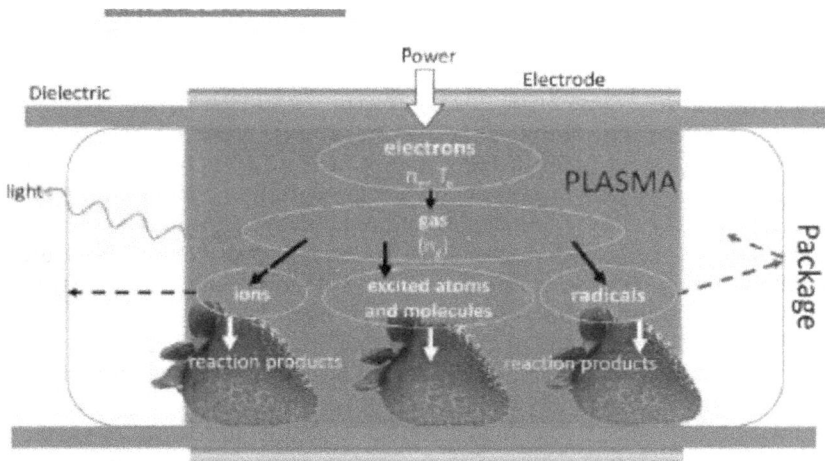

Figure 7.8 In-package plasma treatment of strawberries.

Source: Adapted from Misra et al. (2014)

Misra et al. (2014) did in-package treatment of strawberries for degradation of pesticide residues, showing the effect of plasma on pesticide removal. Figure 7.8 shows in-package treatment of strawberries using cold plasma technology. (Sarangapani et al., 2017) also reported degradation of pesticide residues on blueberries by employing an atmospheric air plasma treatment for 5 minutes, resulting in a reduction of more than 70% of pesticide residues. Also, degradation of diazinon insecticide on cucumber was studied by Dorraki et al. (2016), who reported that exposure of cucumber to plasma treatment from 0.15 to 0.7 W enhanced pesticide degradation efficiency.

Pesticide elimination from the surface of apples using atmospheric plasma was reported by Sarangpani et al., 2017, who stated that exposure to cold plasma led to 95% reduction of paraoxon pesticide along with complete degradation of the microbial population within 2 minutes of exposure. In the grain industry, elimination of toxic fumigation residues has been ascribed to the use of gas-phase plasma treatment besides being used as a control measure for insects like Indian meal moth, mealy bugs and *Tribolium castaneum* (Abd El-Aziz et al., 2014; Ramanan et al., 2018; Ten Bosch et al., 2017).

7.6 Effect of Cold Plasma on Quality Parameters of Different Food Products

Although the promising behaviour of cold plasma in food decontamination is well established, there have also been various reports about quality alteration of foods. Some of the undesirable quality changes that are associated with plasma treatment are surface oxidation, foul odour development and surface discolouration. Changes in ascorbic acid content in foods like lettuce (Song et al., 2015), strawberries (Misra et al., 2014) and blueberries (Sarangapani et al., 2017) have been reported by exposure to cold plasma at a higher dosages approximating 60 to 80 kV. Dielectric barrier discharge plasma sources result in negligible change in the antioxidant capacity of fresh-cut apples and red chicory (Pasquali et al., 2016; Ramazzina et al., 2015). Pigments are an important parameter regarding the apparent quality and antioxidant profile of food products. Exposure to cold plasma treatment has been associated with the degradation of natural pigments like chlorophyll due to the oxidation reaction caused by free radicals, such as chlorophyll degradation in fresh-cut kiwis (Ramazzina et al., 2015). Various studies have been done regarding quality change of fruits after exposure to plasma treatment. Changes in texture of blueberries and cherry tomatoes have revealed a decrease in firmness of blueberries Lacombe et al., 2015), while no significant effect on firmness was found in cherry tomatoes (Misra et al., 2014). On the positive side, desirable changes in the case of phenolic compounds and anthocyanin in sour cherry marasca and pomegranate juice occurred due to short exposure to plasma treatment when compared to conventional thermal treatment processes (Herceg et al., 2016), attributing the enhanced phenolic content to cell disintegration due to plasma treatment. Decontamination of several other food products like meat and milk resulted in a reduction of $E.\ coli$ by 54% in meat and a more than threefold log reduction in milk without any undesirable change in the colour of plasma-treated meat and milk. However, a few contradictory studies related to colour degradation of chicken breasts have been reported by several researchers, with findings of a decrease in redness (a^*) and lightness values (L^*) (Jayasena et al., 2015; Gurol et al., 2012; Rød et al., 2012) after exposure to plasma treatment for 10 minutes. Similarly, an increase in the green value of fresh pork due to cold plasma treatment has been ascribed to the reaction between myoglobin and hydrogen peroxide, resulting in the formation of choleglobin (Misra and Jo, 2017). Pork and beef loin samples subjected to plasma treatment for a longer time displayed an increase in 2-thiobarbituric acid reactive substances (TBARS) (Jayasena et al., 2015; Kim et al., 2015), and an increase in lipid oxidation was also reported when treated by plasma for 10 min (Kim et al., 2011)

Changes in the overall quality characteristics of whole milk on exposure to plasma treatment have been evaluated by many scientists, reporting that plasma treatment leads to noticeable colour change in milk as well as increasing the acidity (Kim et al., 2015). Korachi et al. (2015) investigated biochemical changes in milk, reporting an increase in aldehyde content in milk ascribed to changes caused by the reactive oxygen species; however, no changes were noticed in the fatty acid profile. In a study by Sharma (2020), exposure of raw milk to plasma treatment led to various changes as revealed by Fourier-transform infrared spectroscopy (FTIR), Ultra-performance Liquid Chromatography-tandem Mass Spectrometry (UPLC-MS/MS) and Enzyme linked immuno sorbent assay (ELISA). A decrease in amino acid content, the antigenicity of α-lactalbumin and increase in β-lactoglobulin antigenicity was also observed, thereby paving the way for mitigation of milk allergens during processing. A slight reduction of pH in milk and no effect in pH of egg white was observed after exposure to plasma treatment (Kim et al., 2015; Ragni et al., 2010).

Cold plasma treatment also affects the dough rheology of wheat flour, increasing its viscous and elastic modulus Misra et al., 2015. In addition, lowering of cooking time and enhancement of cooking quality of basmati rice after exposure to cold plasma treatment was reported by Thirumdas et al. (2015).

In addition to physical and chemical changes in food products after exposure to plasma treatment, sensory quality also needs to be addressed. Considering sensory changes in various food products, several studies have revealed that slight changes in flavour and taste of milk, cheese slices and pork loin (Kim et al., 2015) were found, ascribed to lipid oxidation and reactions caused by reactive oxygen species as well as production of alkanes, alkenes, aldehydes, alcohols, ketones and acids (Jayasena et al., 2015).

7.7 Limitations of Cold Plasma Technology

Every coin has two sides; similarly, cold plasma, though a highly acclaimed and constructively useful processing technology, still has a few challenges that limit its scale-up in most sectors in the food industry. The major restrictions associated with commercial scale-up of cold plasma technology are its unstipulated operating parameters and unavailability of enough literature related to the quality characteristics of food (Mandal et al., 2018). The major drawback of cold plasma technology is the occurrence of oxidative changes in some food products containing fats (Pasquali et al., 2016). In addition, some undesirable sensory changes, as discussed in the previous sections, are also one of the major drawbacks of cold plasma exposure of foods. Loss of antioxidants in tomatoes (anthocyanins), cucumber (chlorophyll) and pear and apples and ascorbic acid and colour degradation also pose a bottleneck for the extensive industrial usage of cold plasma technology in the fruit and vegetable industry (Baier et al., 2015).

7.8 Future Scope

Drawbacks associated with the wide scale-up of plasma technology call for an extensive understanding of various factors related to this technique, including free radicals, reactive oxygen species and optimized modes of action regarding inactivation of various microbes and degradation of pertinent residues. Challenges related to undesirable sensory changes on organoleptic qualities and nutritive values also need to be extensively addressed. In addition, widespread research needs to be done reading the toxic or chemical residual remains at the end of the technique. Due to ambiguous regulatory approval, cold plasma is yet to be accepted in the wide public domain, thereby limiting its usage and establishment in food processing. Major parameters related to plasma technology, including the type of gas used and its flow rate, power intensity and voltage, need to be properly standardized in order to scale up the process significantly at the commercial level. Exhaustive capital investment related to the noble gases (Ar, He, etc.) also needs to be addressed so that the process can be economically more feasible and viable. Different sources of plasma generation like microwave-assisted plasma generation, radiofrequency-driven processes and magnetically guided gliding arcs also need to be widely investigated in the food processing sector. One of the major areas of future research in this technique is its usage in the treatment of liquid foods that are minimally explored so far. Cold plasma technology can also be explored as an assisting technique for the enhancement of the overall safety of food products and can be used as an assisting process in hurdle technology. Considering all the areas that can be explored, the future of cold plasma technology seems to be quite positive, which can ultimately pave the way for a broad scale-up of the technique.

7.9 Conclusion

Disinfection of food products by employing cold plasma is still in its infancy; however, considerable literature is available regarding its effectiveness in the food sector. The promising future of this technology regarding food decontamination, package disinfection, waste treatment and pesticide degradation needs to be further explored and investigated thoroughly for commercial scale. Also, the mode of action of plasma treatment for deactivation of microbial cells, enzymes and modification of food functionality is still ambiguous and needs further exploration. Despite extensive research regarding antimicrobial effects of this technique, it is imperative to consider it in mutual agreement with the type of contamination occurring in food products and the processing techniques as well as the end results regarding the healthcare implications and the conditions to explain the optimization of the technical parameters and make it an efficient alternative for food decontamination processes.

References

Abd El-Aziz, M. F., Mahmoud, E. A., & Elaragi, G. M. (2014). Non thermal plasma for control of the Indian meal moth, Plodia interpunctella (Lepidoptera: Pyralidae). *Journal of Stored Products Research, 59,* 215–221.

Ahlfeld, B., Li, Y., Boulaaba, A., Binder, A., Schotte, U., Zimmermann, J. L., . . . Klein, G. (2015). Inactivation of a foodborne norovirus outbreak strain with nonthermal atmospheric pressure plasma. *MBio, 6*(1), e02300–e02314.

Alshraiedeh, N. H., Alkawareek, M. Y., Gorman, S. P., Graham, W. G., & Gilmore, B. F. (2013). Atmospheric pressure, nonthermal plasma inactivation of MS 2 bacteriophage: Effect of oxygen concentration on virucidal activity. *Journal of Applied Microbiology, 115*(6), 1420–1426.

Arjunan, K. P., Sharma, V. K., & Ptasinska, S. (2015). Effects of atmospheric pressure plasmas on isolated and cellular DNA—a review. *International Journal of Molecular Sciences, 16*(2), 2971–3016.

Baier, M., Ehlbeck, J., Knorr, D., Herppich, W. B., & Schlüter, O. (2015). Impact of plasma processed air (PPA) on quality parameters of fresh produce. *Postharvest Biology and Technology, 100,* 120–126.

Bourke, P., Ziuzina, D., Boehm, D., Cullen, P. J., & Keener, K. (2018). The potential of cold plasma for safe and sustainable food production. *Trends in Biotechnology, 36*(6), 615–626.

Charoux, C. M. G., Patange, A., Lamba, S., O'Donnell, C. P., Tiwari, B. K., & Scannell, A. G. M. (2021). Applications of nonthermal plasma technology on safety and quality of dried food ingredients. *Journal of Applied Microbiology, 130*(2), 325–340.

Coutinho, N. M., Silveira, M. R., Rocha, R. S., Moraes, J., Ferreira, M. V. S., Pimentel, T. C., . . . Cruz, A. G. (2018). Cold plasma processing of milk and dairy products. *Trends in Food Science & Technology, 74,* 56–68.

Dasan, B. G., Boyaci, I. H., & Mutlu, M. (2016). Inactivation of aflatoxigenic fungi (Aspergillus spp.) on granular food model, maize, in an atmospheric pressure fluidized bed plasma system. *Food Control, 70,* 1–8.

Devi, Y., Thirumdas, R., Sarangapani, C., Deshmukh, R. R., & Annapure, U. S. (2017). Influence of cold plasma on fungal growth and aflatoxins production on groundnuts. *Food Control, 77,* 187–191.

Dorraki, N., Mahdavi, V., Ghomi, H., & Ghasempour, A. (2016). Elimination of diazinon insecticide from cucumber surface by atmospheric pressure air-dielectric barrier discharge plasma. *Biointerphases, 11*(4), 041007.

Eisenmenger, M. J., & Reyes-De-Corcuera, J. I. (2009). High hydrostatic pressure increased stability and activity of immobilized lipase in hexane. *Enzyme and Microbial Technology, 45*(2), 118–125.

Ekezie, F. G. C., Sun, D. W., & Cheng, J. H. (2017). A review on recent advances in cold plasma technology for the food industry: Current applications and future trends. *Trends in Food Science & Technology, 69*, 46–58.

Fricke, K., Koban, I., Tresp, H., Jablonowski, L., Schröder, K., Kramer, A., . . . Kocher, T. (2012). Atmospheric pressure plasma: A high-performance tool for the efficient removal of biofilms e42539.

Gavahian, M., & Cullen, P. J. (2020). Cold plasma as an emerging technique for mycotoxin-free food: Efficacy, mechanisms, and trends. *Food Reviews International, 36*(2), 193–214.

Gavahian, M., & Khaneghah, A. M. (2020). Cold plasma as a tool for the elimination of food contaminants: Recent advances and future trends. *Critical Reviews in Food Science and Nutrition, 60*(9), 1581–1592.

Gavahian, M., Pallares, N., Al Khawli, F., Ferrer, E., & Barba, F. J. (2020). Recent advances in the application of innovative food processing technologies for mycotoxins and pesticide reduction in foods. *Trends in Food Science & Technology, 106*, 209–218.

Gavahian, M., Peng, H. J., & Chu, Y. H. (2019). Efficacy of cold plasma in producing Salmonella-free duck eggs: Effects on physical characteristics, lipid oxidation, and fatty acid profile. *Journal of Food Science and Technology, 56*(12), 5271–5281.

Gurol, C., Ekinci, F. Y., Aslan, N., & Korachi, M. (2012). Low temperature plasma for decontamination of E. coli in milk. *International Journal of Food Microbiology, 157*(1), 1–5.

Han, L., Boehm, D., Amias, E., Milosavljević, V., Cullen, P. J., & Bourke, P. (2016). Atmospheric cold plasma interactions with modified atmosphere packaging inducer gases for safe food preservation. *Innovative Food Science & Emerging Technologies, 38*, 384–392.

Herceg, Z., Kovačević, D. B., Kljusurić, J. G., Jambrak, A. R., Zorić, Z., & Dragović-Uzelac, V. (2016). Gas phase plasma impact on phenolic compounds in pomegranate juice. *Food Chemistry, 190*, 665–672.

Herron, J. T., & Green, D. S. (2001). Chemical kinetics database and predictive schemes for nonthermal humid air plasma chemistry. Part II. Neutral species reactions. *Plasma Chemistry and Plasma Processing, 21*(3), 459–481.

Hertwig, C., Leslie, A., Meneses, N., Reineke, K., Rauh, C., & Schlüter, O. (2017). Inactivation of Salmonella Enteritidis PT30 on the surface of unpeeled almonds by cold plasma. *Innovative Food Science & Emerging Technologies, 44*, 242–248.

Hertwig, C., Reineke, K., Ehlbeck, J., Knorr, D., & Schlüter, O. (2015). Decontamination of whole black pepper using different cold atmospheric pressure plasma applications. *Food Control, 55*, 221–229.

Hojnik, N., Cvelbar, U., Tavčar-Kalcher, G., Walsh, J. L., & Križaj, I. (2017). Mycotoxin decontamination of food: Cold atmospheric pressure plasma versus "classic" decontamination. *Toxins, 9*(5), 151.

Jayasena, D. D., Kim, H. J., Yong, H. I., Park, S., Kim, K., Choe, W., & Jo, C. (2015). Flexible thin-layer dielectric barrier discharge plasma treatment of pork butt and beef loin: Effects on pathogen inactivation and meat-quality attributes. *Food Microbiology, 46*, 51–57.

Joshi, S. G., Cooper, M., Yost, A., Paff, M., Ercan, U. K., Fridman, G., . . . Brooks, A. D. (2011). Nonthermal dielectric-barrier discharge plasma-induced inactivation involves oxidative DNA damage and membrane lipid peroxidation in Escherichia coli. *Antimicrobial Agents and Chemotherapy, 55*(3), 1053–1062.

Jouany, J. P. (2007). Methods for preventing, decontaminating and minimizing the toxicity of mycotoxins in feeds. *Animal Feed Science and Technology, 137*(3–4), 342–362.

Kashfi, A. S., Ramezan, Y., & Khani, M. R. (2020). Simultaneous study of the antioxidant activity, microbial decontamination and color of dried peppermint (Mentha piperita L.) using low pressure cold plasma. *LWT, 123*, 109121.

Khaneghah, A. M., Hashemi, S. M. B., & Limbo, S. (2018). Antimicrobial agents and packaging systems in antimicrobial active food packaging: An overview of approaches and interactions. *Food and Bioproducts Processing, 111*, 1–19.

Kim, D. H., Je, Y. J., Kim, C. D., Lee, Y. H., Seo, Y. J., Lee, J. H., & Lee, Y. (2011). Can platelet-rich plasma be used for skin rejuvenation? Evaluation of effects of platelet-rich plasma on human dermal fibroblast. *Annals of Dermatology, 23*(4), 424–431.

Kim, H. J., Yong, H. I., Park, S., Kim, K., Choe, W., & Jo, C. (2015). Microbial safety and quality attributes of milk following treatment with atmospheric pressure encapsulated dielectric barrier discharge plasma. *Food Control, 47*, 451–456.

Kim, J. H., & Min, S. C. (2017). Microwave-powered cold plasma treatment for improving microbiological safety of cherry tomato against Salmonella. *Postharvest Biology and Technology, 127*, 21–26.

Korachi, M., Ozen, F., Aslan, N., Vannini, L., Guerzoni, M. E., Gottardi, D., & Ekinci, F. Y. (2015). Biochemical changes to milk following treatment by a novel, cold atmospheric plasma system. *International Dairy Journal, 42*, 64–69.

Koutchma, T. (2009). Advances in ultraviolet light technology for non-thermal processing of liquid foods. *Food and Bioprocess Technology, 2*(2), 138–155.

Kuan, Y. H., Bhat, R., Patras, A., & Karim, A. A. (2013). Radiation processing of food proteins–A review on the recent developments. *Trends in Food Science & Technology, 30*(2), 105–120.

Lacombe, A., Niemira, B. A., Gurtler, J. B., Fan, X., Sites, J., Boyd, G., & Chen, H. (2015). Atmospheric cold plasma inactivation of aerobic microorganisms on blueberries and effects on quality attributes. *Food Microbiology, 46*, 479–484.

Leins, M., Kopecki, J., Gaiser, S., Schulz, A., Walker, M., Schumacher, U., . . . Hirth, T. (2014). Microwave plasmas at atmospheric pressure. *Contributions to Plasma Physics, 54*(1), 14–26.

Leipold, F., Kusano, Y., Hansen, F., & Jacobsen, T. (2010). Decontamination of a rotating cutting tool during operation by means of atmospheric pressure plasmas. *Food Control, 21*(8), 1194–1198.

Lu, X., Blawert, C., Kainer, K. U., & Zheludkevich, M. L. (2016). Investigation of the formation mechanisms of plasma electrolytic oxidation coatings on Mg alloy AM50 using particles. *Electrochimica Acta, 196*, 680–691.

Mai-Prochnow, A., Murphy, A. B., McLean, K. M., Kong, M. G., & Ostrikov, K. K. (2014). Atmospheric pressure plasmas: Infection control and bacterial responses. *International Journal of Antimicrobial Agents, 43*(6), 508–517.

Mandal, R., Singh, A., & Singh, A. P. (2018). Recent developments in cold plasma decontamination technology in the food industry. *Trends in Food Science & Technology, 80*, 93–103.

Mendis, D. A., Rosenberg, M., & Azam, F. (2000). A note on the possible electrostatic disruption of bacteria. *IEEE Transactions on Plasma Science, 28*(4), 1304–1306.

Misra, N. N., & Jo, C. (2017). Applications of cold plasma technology for microbiological safety in meat industry. *Trends in Food Science & Technology, 64*, 74–86.

Misra, N. N., Martynenko, A., Chemat, F., Paniwnyk, L., Barba, F. J., & Jambrak, A. R. (2018). Thermodynamics, transport phenomena, and electrochemistry of external field-assisted nonthermal food technologies. *Critical Reviews in Food Science and Nutrition, 58*(11), 1832–1863.

Misra, N. N., Pankaj, S. K., Frias, J. M., Keener, K. M., & Cullen, P. J. (2015). The effects of nonthermal plasma on chemical quality of strawberries. *Postharvest Biology and Technology, 110*, 197–202.

Misra, N. N., Pankaj, S. K., Segat, A., & Ishikawa, K. (2016). Cold plasma interactions with enzymes in foods and model systems. *Trends in Food Science & Technology, 55*, 39–47.

Misra, N. N., Patil, S., Moiseev, T., Bourke, P., Mosnier, J. P., Keener, K. M., & Cullen, P. J. (2014). In-package atmospheric pressure cold plasma treatment of strawberries. *Journal of Food Engineering, 125*, 131–138.

Misra, N. N., & Roopesh, M. S. (2019). Cold plasma for sustainable food production and processing. In *Green food processing techniques* (pp. 431–453). Academic Press.

Misra, N. N., Yepez, X., Xu, L., & Keener, K. (2019). In-package cold plasma technologies. *Journal of Food Engineering, 244*, 21–31.

Moreau, M., Orange, N., & Feuilloley, M. G. J. (2008). Non-thermal plasma technologies: New tools for bio-decontamination. *Biotechnology Advances, 26*(6), 610–617.

Moutiq, R., Misra, N. N., Mendonça, A., & Keener, K. (2020). In-package decontamination of chicken breast using cold plasma technology: Microbial, quality and storage studies. *Meat Science, 159*, 107942.

O'Donnell, C. P., Tiwari, B. K., Bourke, P., & Cullen, P. J. (2010). Effect of ultrasonic processing on food enzymes of industrial importance. *Trends in Food Science & Technology, 21*(7), 358–367.

Pankaj, S. K., Bueno-Ferrer, C., Misra, N. N., O'Neill, L., Jiménez, A., Bourke, P., & Cullen, P. J. (2014). Characterization of polylactic acid films for food packaging as affected by dielectric barrier discharge atmospheric plasma. *Innovative Food Science & Emerging Technologies, 21*, 107–113.

Pankaj, S. K., Misra, N. N., & Cullen, P. J. (2013). Kinetics of tomato peroxidase inactivation by atmospheric pressure cold plasma based on dielectric barrier discharge. *Innovative Food Science & Emerging Technologies, 19*, 153–157.

Pankaj, S. K., Shi, H., & Keener, K. M. (2018). A review of novel physical and chemical decontamination technologies for aflatoxin in food. *Trends in Food Science & Technology, 71*, 73–83.

Park, S. Y., & Ha, S. D. (2015). Application of cold oxygen plasma for the reduction of Cladosporium cladosporioides and Penicillium citrinum on the surface of dried filefish (Stephanolepis cirrhifer) fillets. *International Journal of Food Science & Technology, 50*(4), 966–973.

Pasquali, F., Stratakos, A. C., Koidis, A., Berardinelli, A., Cevoli, C., Ragni, L., . . . & Trevisani, M. (2016). Atmospheric cold plasma process for vegetable leaf decontamination: A feasibility study on radicchio (red chicory, Cichorium intybus L.). *Food Control, 60*, 552–559.

Pignata, C., D'angelo, D., Fea, E., & Gilli, G. (2017). A review on microbiological decontamination of fresh produce with nonthermal plasma. *Journal of Applied Microbiology, 122*(6), 1438–1455.

Ragni, L., Berardinelli, A., Vannini, L., Montanari, C., Sirri, F., Guerzoni, M. E., & Guarnieri, A. (2010). Non-thermal atmospheric gas plasma device for surface decontamination of shell eggs. *Journal of Food Engineering, 100*(1), 125–132.

Ramanan, K. R., Sarumathi, R., & Mahendran, R. (2018). Influence of cold plasma on mortality rate of different life stages of Tribolium castaneum on refined wheat flour. *Journal of Stored Products Research, 77*, 126–134.

Ramazzina, I., Berardinelli, A., Rizzi, F., Tappi, S., Ragni, L., Sacchetti, G., & Rocculi, P. (2015). Effect of cold plasma treatment on physico-chemical parameters and antioxidant activity of minimally processed kiwi-fruit. *Postharvest Biology and Technology, 107,* 55–65.

Randeniya, L. K., & de Groot, G. J. (2015). Non-thermal plasma treatment of agricultural seeds for stimulation of germination, removal of surface contamination and other benefits: A review. *Plasma Processes and Polymers, 12*(7), 608–623.

Razzaghi, N., Ziarati, P., Rastegar, H., Shoeibi, S., Amirahmadi, M., Conti, G. O., . . . & Khaneghah, A. M. (2018). The concentration and probabilistic health risk assessment of pesticide residues in commercially available olive oils in Iran. *Food and chemical toxicology, 120,* 32–40.

Rød, S. K., Hansen, F., Leipold, F., & Knøchel, S. (2012). Cold atmospheric pressure plasma treatment of ready-to-eat meat: Inactivation of Listeria innocua and changes in product quality. *Food Microbiology, 30*(1), 233–238.

Sarangapani, C., Danaher, M., Tiwari, B., Lu, P., Bourke, P., & Cullen, P. J. (2017). Efficacy and mechanistic insights into endocrine disruptor degradation using atmospheric air plasma. *Chemical Engineering Journal, 326,* 700–714.

Sarangapani, C., Lu, P., Behan, P., Bourke, P., & Cullen, P. J. (2018). Humic acid and trihalomethane breakdown with potential by-product formations for atmospheric air plasma water treatment. *Journal of Industrial and Engineering Chemistry, 59,* 350–361.

Sarangapani, C., Thirumdas, R., Devi, Y., Trimukhe, A., Deshmukh, R. R., & Annapure, U. S. (2016). Effect of low-pressure plasma on physico-chemical and functional properties of parboiled rice flour. *LWT-Food Science and Technology, 69,* 482–489.

Schaarschmidt, S. (2016). Public and private standards for dried culinary herbs and spices—part I: Standards defining the physical and chemical product quality and safety. *Food Control, 70,* 339–349.

Sharma, S. (2020). Cold plasma treatment of dairy proteins in relation to functionality enhancement. *Trends in Food Science & Technology, 102,* 30–36.

Shi, H., Ileleji, K., Stroshine, R. L., Keener, K., & Jensen, J. L. (2017). Reduction of aflatoxin in corn by high voltage atmospheric cold plasma. *Food and Bioprocess Technology, 10*(6), 1042–1052.

Sicherer, S. H., & Sampson, H. A. (2018). Food allergy: A review and update on epidemiology, pathogenesis, diagnosis, prevention, and management. *Journal of Allergy and Clinical Immunology, 141*(1), 41–58.

Song, T., Berrehrah, H., Cabrera, D., Torres-Rincon, J. M., Tolos, L., Cassing, W., & Bratkovskaya, E. (2015). Tomography of the quark-gluon plasma by charm quarks. *Physical Review C, 92*(1), 014910.

Stoffels, E., Sakiyama, Y., & Graves, D. B. (2008). Cold atmospheric plasma: Charged species and their interactions with cells and tissues. *IEEE Transactions on Plasma Science, 36*(4), 1441–1457.

Suhem, K., Matan, N., Nisoa, M., & Matan, N. (2013). Inhibition of Aspergillus flavus on agar media and brown rice cereal bars using cold atmospheric plasma treatment. *International Journal of Food Microbiology*, *161*(2), 107–111.

Ten Bosch, L., Pfohl, K., Avramidis, G., Wieneke, S., Viöl, W., & Karlovsky, P. (2017). Plasma-based degradation of mycotoxins produced by Fusarium, Aspergillus and Alternaria species. *Toxins*, *9*(3), 97.

Thirumdas, R., Sarangapani, C., & Annapure, U. S. (2015). Cold plasma: a novel non-thermal technology for food processing. *Food biophysics*, 10, 1–11.

Ulbin-Figlewicz, N., Brychcy, E., & Jarmoluk, A. (2015). Effect of low-pressure cold plasma on surface microflora of meat and quality attributes. *Journal of Food Science and Technology*, *52*(2), 1228–1232.

Van Gaens, W., & Bogaerts, A. (2013). Kinetic modelling for an atmospheric pressure argon plasma jet in humid air. *Journal of Physics D: Applied Physics*, *46*(27), 275201.

Varilla, C., Marcone, M., & Annor, G. A. (2020). Potential of cold plasma technology in ensuring the safety of foods and agricultural produce: A review. *Foods*, *9*(10), 1435.

Venezia, R. A., Orrico, M., Houston, E., Yin, S. M., & Naumova, Y. Y. (2008). Lethal activity of nonthermal plasma sterilization against microorganisms. *Infection Control & Hospital Epidemiology*, *29*(5), 430–436.

Wang, X., Wang, S., Yan, Y., Wang, W., Zhang, L., & Zong, W. (2020). The degradation of Alternaria mycotoxins by dielectric barrier discharge cold plasma. *Food Control*, *117*, 107333.

Wielogorska, E., Ahmed, Y., Meneely, J., Graham, W. G., Elliott, C. T., & Gilmore, B. F. (2019). A holistic study to understand the detoxification of mycotoxins in maize and impact on its molecular integrity using cold atmospheric plasma treatment. *Food Chemistry*, *301*, 125281.

Yong, H. I., Lee, H., Park, S., Park, J., Choe, W., Jung, S., & Jo, C. (2017). Flexible thin-layer plasma inactivation of bacteria and mold survival in beef jerky packaging and its effects on the meat's physicochemical properties. *Meat Science*, *123*, 151–156.

Yousefi, M., Ghoochani, M., & Mahvi, A. H. (2018). Health risk assessment to fluoride in drinking water of rural residents living in the Poldasht city, Northwest of Iran. *Ecotoxicology and Environmental Safety*, *148*, 426–430.

Yusupov, M., Neyts, E. C., Khalilov, U., Snoeckx, R., Van Duin, A. C. T., & Bogaerts, A. (2012). Atomic-scale simulations of reactive oxygen plasma species interacting with bacterial cell walls. *New Journal of Physics*, *14*(9), 093043.

Zhang, H., Ma, D., Qiu, R., Tang, Y., & Du, C. (2017). Non-thermal plasma technology for organic contaminated soil remediation: A review. *Chemical Engineering Journal*, *313*, 157–170.

Zhao, W., Yang, R., & Zhang, H. Q. (2012). Recent advances in the action of pulsed electric fields on enzymes and food component proteins. *Trends in Food Science & Technology, 27*(2), 83–96.

Zhuang, H., Rothrock Jr, M. J., Line, J. E., Lawrence, K. C., Gamble, G. R., Bowker, B. C., & Keener, K. M. (2020). Optimization of in-package cold plasma treatment conditions for raw chicken breast meat with response surface methodology. *Innovative Food Science & Emerging Technologies, 66*, 102477.

Chapter 8

Biomimicry– An Approach to Food Product and Technology Innovation

Rajan Sharma, Gurkirat Kaur, and Savita Sharma

8.1 Introduction

Since life appeared on Earth, nature has gone through evolution, with excellent performance from the nano scale to the macro scale. The perfection in the functionality of natural objects and processes has attracted interest from researchers and innovators to derive and adapt biologically inspired designs, referred to as "biomimetics", which signifies mimicry of natural processes, their principles and their biology (Bhushan, 2009). The term is taken from the Greek word "biomimesis" and was first used in 1957 doctoral candidate Otto Schmitt, who adapted the electrical action of a nerve to develop a physical device. It has been argued that natural designs are superior in comparison to human creative thinking in several domains; thus mimicking their characteristic features and basic principles could result in the improvement of existing human-made technology (Bar-Cohen, 2006). Most natural objects and processes have cell-based functionality, which offers the unique advantages of self-repair and fault tolerance during regular biological activities. Therefore, success in the development of

DOI: 10.1201/9781003217138-8

biomimetic objects could bring innovative results, bringing science fiction to reality. Historians have argued that our ancestors understood biological processes in nature as a real source of inspiration from the beginning of science (Aziz, 2016). Authors have further documented that the 20th century was the period of physics (engineering), while the 21st century is the period of biology (life sciences). Life sciences are now bigger than engineering in terms of research funding, significant discoveries and size of the workforce.

The available literature on biomimetic objects and processes suggests that there could be three levels of biomimicry to address the transformation of existing technologies (Aziz, 2016):

1. The organism level: It involves imitating the principal phenomenon of certain organisms or impersonating a fragment of a whole organism.
2. The behavior level: It is the simulation of the behavior of the organism to design a newer concept for the modern world.
3. The ecosystem level: It is the most challenging level, as it refers to mimicry of whole ecosystems.

There are five different dimensions to characterize targeted biomimetic solutions: form (to mimic the look of the natural object or process), material (to simulate the compositional character), construction (to replicate biological management), process (to mimic the biological working of any object) and function (to imitate the final outcome of any natural object or process) (Aziz, 2016).

Adaptation of a natural phenomenon to create novelty in artificial objects has been seen in various disciplines, including material science, robotics, architecture and computer science (Lepora et al., 2013). More recently, this concept has driven a novel research agenda in food industries by designing food products and processing operations on the underlying principles of natural functions. Among many other, the following are the major achievements of the food industry in designing food products and technologies with the inspiration of biological processes:

- **Cultured meat:** It is the imitation of meat products from *in vitro* cultivation of animal cells instead of slaughtering animals.
- **Biomimetic membranes:** These are food processing aids that mimic or incorporate biological elements or ideas. Transport efficiency and functionality of existing systems can be improved using strategies evolved by nature.
- **Biomimetic milk:** It has been reported that lipid structure in human milk modulates digestion kinetics and is involved in metabolic programming. Thus, biomimetic milk is the simulation of

human milk in terms of biochemical composition and structural features, as both these factors play significant roles in its nutritive potential.

- **Bioartificial tongues:** They have been designed to mimic the taste-sensing properties of the human tongue. It is a significant approach in objective organoleptic evaluation of food products and is an *in vitro* model for elucidation of the mechanisms of taste.
- **Active packaging:** Active packaging refers to additional functions of preservation of food beyond physical containment and protection. They may include oxygen scavengers or ethylene scavengers.

8.2 Cultured Meat

Over the past few years, there has been an increase in research investments in alternative meat sources to conventional meat from livestock. The need has arisen due to three predominant factors, as described by Post (2012):

1. There has been a tremendous increase in the demand for meat products, and the current system will run out of stock in the near future, as already livestock feeding and management constitute a significant portion of arable land.
2. The associated environmental impact of breeding and management of livestock is one of the major concerns around the globe.
3. Animal welfare and public health concerns are also rising with an increase in the herding volume and slaughtering capacity.

One promising meat alternative is the culturing of meat. Cultured meat, also referred to as lab-grown, synthetic and *in vitro* meat, has been defined as meat developed in a bioreactor using biochemical engineering principles (Zhang et al., 2020). It is among the best solutions for those who are involved in animal welfare but at the same time do not wish to change the composition of their diet. The idea of *in vitro* production of meat developed from advancements in tissue engineering in the world of regenerative medicine, where muscle tissues could be developed from stem cells (Arshad et al., 2017). Dr. Mark Post, a pharmacologist, developed a prototype of a cultured meat patty in 2013, which had similar organoleptic characteristics as conventional meat; however, it took more than $300,000 and a period of about three months to culture a meat patty of around 5 ounces. Cultured meat could not only address the need for a conventional meat replica but could also improve the compositional and functional attributes of conventional meat. For example, stem cells for the production of cultured meat could be taken from any source or blend of sources, diversifying the scope to unimaginable products. Meats could be customized according to the needs of the consumer or designed for specific health conditions (Post, 2012).

8.2.1 Stem Cells—The Preferred Source

Conventional meat consists of skeletal muscles, including fibroblasts, endothelial cells, adipocytes and leucocytes, which are the major contributors to its palatability in terms of texture and flavor. The preferable source cells for the biochemical engineering of skeletal muscles are myoblast to satellite stem cells, as:

- They are involved in the regeneration of the muscles after injury (Mauro, 1961)
- Although they do not easily maintain their replicative state in the culturing medium, once culturing has reached a sufficient number, they can easily differentiate to develop more myotubes and myofibrils (Post, 2012)
- Recent findings suggest that there may be some subsets of these satellite stem cells which possess even better regenerative potential (Collins et al., 2007)
- Ascribed to their ability to repair damages and diseased cells, they show self-renewal properties, and consequently, they can be differentiated and transformed into different types of cells; for instance, pluripotent adult cells have been assessed for *in vitro* culturing of meat (Slack, 2008)

8.2.2 Culturing of Cells

The most challenging task for the *in vitro* production of meat is the identification of the correct culturing medium for growth and proliferation of stem cells. Bhat and Fayaz (2011) reported that there were two proposals for the production of *in vitro* meat from cell culture, one of which suggested growing targeted cells together with collagen spheres that could act as a scaffold to hold and retain myoblasts, while the other stated a culture medium may be percolated through a meshwork of collagen after regular time intervals, and once cell differentiation started, the resultant product made of collagen and muscle cells could be harvested and confirmed as cultured meat.

It is also important to standardize the composition of culturing media. Supplementation of the medium with serum has been found to accelerate the growth of cells; however, there may be some detrimental effects, as it is not an actual biological component of pathological fluid. Ultoser G is a commercially available serum substitute which contains about 20% of total serum protein along with vitamins, hormones, growth factors, minerals and binding proteins. The growth of cells has been reported as comparatively higher than that of serum. Culturing of cells has been broadly separated into two distinct phases (Post, 2012):

1. **The proliferation phase:** It aims to maximize the starting batch of cells so as to achieve a higher rate of doubling. Delay in the differentiation phase could also help increase the doubling numbers. Further, the replicative ability of satellite cells has been reported to be enhanced if skeletal muscle fibrils are exposed to trituration during harvesting and with the use of enzymatic treatment.
2. **The differentiation phase:** Once sufficient cells are produced, they are targeted for differentiation into skeletal muscle cells. Few changes are needed in the natural culture conditions for the cells under hypertrophy. However, any of mechanical, biochemical and metabolic stimuli could assist in triggering protein synthesis and furthering their organization into contractile muscles. The proteins at this stage start to attain their microscopic morphological features. They could be seen as collagen-like gel fixed on a bio-scaffold.

In general, *in vitro* production of meat through tissue engineering approaches could be explained as three-dimensional branch networking of biological polymers or a scaffold containing the necessary nutrients to which satellite cells or myoblasts could attach (Zandonella, 2003). The most commonly used scaffolds are microcarrier beads and collagen-based meshwork due to their biodegradability and biocompatibility. When these cells re fused onto the scaffold, it results in the formation of myotubes, and their differentiation produces myofibers; consequently, boneless meat of soft consistency and customized nutritional properties could be achieved (Noor et al., 2016). Moreover, Bhat and Fayaz (2011) stated that tissue culturing is advantageous, as explants in this case comprise all the necessary tissues in accurate proportion, and the resultant product better mimics the *in vivo* situation; however, they further added that the production process can only be controlled to a small extent. *In vitro* muscle tissues contain contractile proteins as a predominant constituent; however, the presence of other proteins could result in quality enhancement of the resultant meat in terms of color, taste and texture. For instance, heme-carrying protein could impart a pink color to the meat, which is a characteristic feature of conventional meat; further, it will add to the taste, being a significant carrier of iron (Kanatous & Mammen, 2010). Potential advantages and key challenges of cultured meat are shown in Figure 8.1.

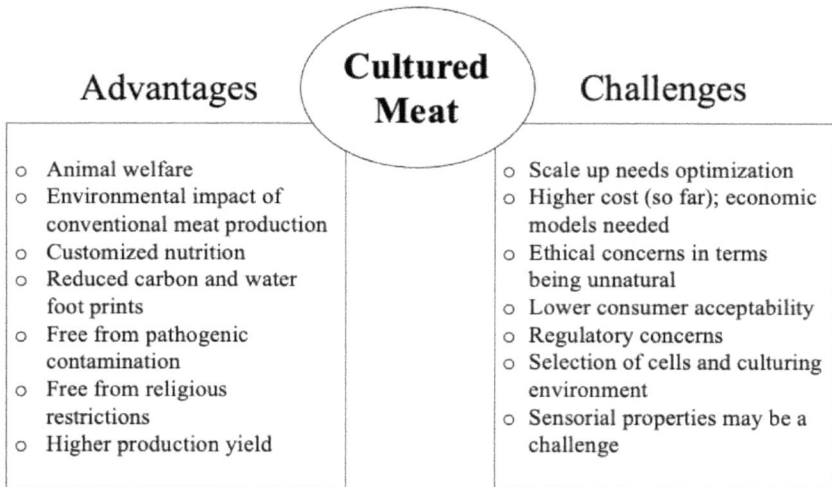

Figure 8.1 Potential advantages and challenges with cultured meat.

8.3 Biomimetic Milk

It is a well-established fact that during early infancy, human infants must be nursed with mother's breast milk, and no artificial infant formula is an exact replacement for breast milk. Further, the World Health Organization recommended that infants should be fed only mothers' milk at least for four to six months after birth (Sarkar, 2003). During the last few years, it has been noted that there is a significant decline in breastfeeding due to lack of awareness, modern lifestyle, urbanization, social status and industrialization (Misra, 1982). Further factors contributing to this trend could include biological factors such as illness of mother or child, lack of mother's interest, insufficient lactation or child's refusal to suckle (Sarkar, 2003). Over the past few years, there have been enormous advancements in the development of infant formulae in terms of replication of the nutritional quality, safety and therapeutic properties of breast milk; however, no such substitute exists to date which possess comparable compositional and biological properties to those of breast milk.

Despite the excellent nutritive and pharmacological potential of breast milk, the need for biomimetic milk has been seen in several cases including (Sarkar, 2003):

- Prolonged breastfeeding poses the risk of transmission of human immunodeficiency virus (HIV-I) (Smith & Kuhn, 2000)

- Transmission of drugs taken by the mother for sterility therapy after birth may cause allergies or other side effects and a high risk of breast milk jaundice (Sarkar, 2003)
- High levels of cholesterol in human milk may elevate the cholesterol level in infants (Kallio et al., 1992)
- The nutritional composition of human milk is not suitable for infants with significantly low birth weight (Sánchez-Hidalgo et al., 2000)

The biochemical composition of human milk has been established since the 1800s, and various products to target infant formulae based on this composition have been developed and introduced in the market. However, over the past decade, it has been observed that it is not just the nutrient composition of breast milk which makes it a unique product, but its structure also has significant biological potential. Despite numerous attempts to optimize the nutritional composition of infant formulae, no product has ever replicated the bioactivity and immune-protective functions of human milk. The significance of lipid structure of milk, which plays a major role in digestion kinetics and absorption of other nutrients, is now understood (Bourlieu et al., 2015, 2016). Lipids in human milk are present as dispersed droplets which are exceptional bio-physical elements distinguished from other lipoproteins due to the presence of a tri-layered membrane. This membrane constitutes several biologically active polar lipids (including sphingolipids, glycosphingolipids and glycerophospholipids), proteins, enzymes and cholesterol in addition to some trace components such as RNA and mRNA (Bourlieu-Lacanal et al., 2017; Lopez et al., 2015; Lopez & Ménard, 2011). Thus, it may be inferred that heterogenous composition of this membrane induces phase co-existence, which is a specific physical state where nanodomains such as cholesterol and sphingolipids which aggregate along the membrane surface are in an ordered state, resulting in high rigidity, while other biochemical constituents with unsaturated molecules are present in a liquid state (Bourlieu-Lacanal et al., 2017; Gallier et al., 2010; Lopez & Ménard, 2011).

The size of milk fat globules in human milk is around 4 μm, while lipid droplets generally found in infant formulae are 0.5 μm, and this difference dictates the overall variation in the emulsion disintegration and digestibility of the milk in the intestinal tract of infants. In order to mimic breast milk, the stability of such large fat molecules is a real technological challenge. The substitution of bovine milk fat globule membrane extracts in infant formulae has been evaluated for their ability to mimic human milk (Hernell et al., 2016). These fractions purified from dairy by-products (Conway et al., 2014), when added to infant formulae as an ingredient, have been reported to give a better ultrastructure which is in close proximity to human milk. Also, the lipase activity during digestion is dependent upon the surface composition of milk fats; thus, the interfacial attributes of the milk fat globule membrane affect the hydrolysis of membrane lipids, and accordingly assimilation of other nutrients is also influenced. The supplementation of milk fat globule membranes in infant formulae was found to show positive outcomes in terms of nutrition, immunity, gut health, behavior and cognition after different experiments as reviewed by Bourlieu-Lacanal et al. (2017). However, such products remain in their infancy, and their complete and systematic characterization for nutritional, biological, functional and bioactive properties needs to be addressed.

Another approach to mimicking human milk could be the simulation of the core of milk fat globules. It is understood that human milk has a characteristic feature of specific regiodistribution of triglycerides, which is different from most studied infant formulae. Martin et al. (1993) stated that this specific regiodistribution of fatty acid on triglycerol remains constant during different stages of lactation and plays a significant role in lipid hydrolysis and its uptake, ascribed to the fact that lipases in the human gut are stereo- and regioselective (Zou et al., 2012). Efforts have been made to replicate the structural and compositional features of triglycerides of human milk, and commercially available examples include fat analogs (Betapol and INFAT). Bourlieu-Lacanal et al. (2017) further documented that these analogs were prepared after biochemical or enzymatic modification so as to imitate the fatty acid conformation in human milk. More recent advancements have been made to enhance the functionality of cultured human milk; for instance, Zeng et al. (2019) formulated a nanoencapsulated suspension from a plant protein source, zein, to mimic the structure of milk with the objective of inducing clinical advantages of supplementation of docosahexaenoic acid (DHA), which otherwise shows no effects when consumed by children in their infant formulae. The authors found significant changes due to nanoencapsulation in the assimilation pattern of DHA, and further, motor and mental skills were improved. The supplemented infants had better memory and learning potential. The underlying principle of mimicking was the impersonation of the milk fat globule membrane with nanoencapsulated DHA, which stabilized the suspension with a hydrophobic–hydrophilic balance of the system (Dong et al., 2013, 2015). The resultant product has similar biological properties to those of human milk and exhibited no toxicity. Recent approaches to developing biomimetic infant formulae are shown in Table 8.1.

8.4 Electronic Tongues

There has been tremendous advancement in food analytical techniques in the recent past; however, the human abilities of taste and smell have not been comprehensively understood. There are thousands of olfactory receptors present on the tongue, which bind flavor molecules, and our brain can

Table 8.1 Recent Approaches to Biomimicry of Human Milk for Potential Infant Formulae Applications

Biomimetic Approach	Key Outcomes	Reference(s)
Biomimetic milk fat globules	Influence of protein-to-phospholipid ratio dictated significant variation in the rheological, interfacial and physicochemical properties of biomimetic milk, affecting its design and potential applications	(Chen & Sagis, 2019)
Nanoencapsulated suspensions to replicate milk structure	Enhancement in maternal and fetal absorption of DHA; resulted in early brain development, as noted for mice; no changes in *in vivo* safety studies of nanoencapsulated suspensions	(Zeng et al., 2019)
Lipid matrix bioinspired by breast milk	Significant reduction in body fat accumulation as tested in mice; larger size of fat globules coated with phospholipid membrane; modulated structure was responsible for physiological functionality	(Oosting et al., 2011)

detect different flavors even when their concentration is as low as parts per million (ppm) (Glatz & Bailey-Hill, 2011). Organoleptic evaluation of food materials is a critical parameter dictating the overall acceptability of products in the market; however, there are certain limitations to the existing methods of sensory evaluation, such as time consumption, cost, bias, sample carryover, panelist fatigue and large number of trained panelists (Ross, 2021). Thus, there has always been a desire for the quantitative estimation of a sensory profile of developed products. Mimicking the human palate has been done with electronic tongues, also known as e-tongues or bioartificial tongues, which are sensor-based electronic devices that aim to detect and quantify the soluble components when immersed in a food material.

The information is sent in the form of signals to a processing system, and their characterization is done on the basis of pattern recognition (Ross, 2021). Electronic noses also work on a similar principle; the only difference is the nature of the sensors. E-noses include gas sensors, while e-tongues take into account chemical sensors for the detection of flavor compounds (Orlandi et al., 2019). These sensors are promising food analytical equipment with relatively low cost in comparison to available techniques such as high-performance liquid chromatography (HPLC), gas chromatography–mass spectrometry (GC–MS) and laser scattering analyzers. They hold significant applications in terms of sensory evaluation, food safety, processing quality and microbiological attributes (Matindoust et al., 2016; Tan & Xu, 2020).

The working of these bio-artificial sensors can be explained with two major operations, a sensing array and pattern recognition.

8.4.1 Sensing Arrays in E-Tongues

The sensor elements which when come in contact with the flavoring compounds, there occurs series of biochemical reactions resulting in the transformation of electrons to generate electrical signals (Shi et al., 2018). In particular, these sensors may be categorized as:

1. **Potentiometry-based sensors:** They measure the difference between the electrode potential of the targeted sample between the reference electrode and the outward sensor membrane boundary. These systems present certain advantages, like rapid detection, analysis of complex food samples, simple measuring setup and reproducibility over others (Jiang et al., 2018). However, the drawbacks include the fact that the adsorption of soluble components could change the charge transfer, and their performance is dependent upon operation temperature (Ciosek & Wróblewski, 2007).

2. **Voltammetry-based sensors:** In this case, the principle of electrochemical voltammetry is used, and the multi-sensor array is exposed to a targeted sample solution which adds to the step potential of the working electrode. Qualitative and quantitative estimations can be done from the polarization currents of different solutions. The working electrode is usually a bare metal or composed of gold, silver, tin, rhodium, platinum, nickel and copper. These sensors are best suited for samples that show reasonable oxidation-reduction potential, such as identification of milk adulterants, monitoring of extent of fermentation and the paper and

pulp industry (Ivarsson et al., 2005; Jiang et al., 2018; Winquist et al., 2002).

3. **Impedance spectroscopy-based sensors:** These are impedance-sensing elements, usually composed of ultrathin layers which work on the principle of electrochemical impedance spectroscopy. These sensors, when they come into contact with sample solutions, result in a fingerprint of the solution. They have been reported to possess experimental simplicity and lower detection time. The major applications of these kinds of sensors include qualitative estimation of compounds in tea, coffee, wine and mineral water (Bhondekar et al., 2011; Jiang et al., 2018). They offer certain advantages like not needing reference electrodes, and their sensing units may not be electroactive (Elamine et al., 2019).

8.4.2 Pattern Recognition in E-Tongues

The processing of signals sent by different sensing elements is carried out in terms of pattern recognition, which are potential statistical techniques capable of automatically classifying responses (Shi et al., 2018). The most commonly used pattern recognition techniques include principal component analysis (PCA), artificial neural networks (ANNs), support vector machines (SVMs), partial least squares (PLS), linear discriminant analysis (LDA), functional discriminant analysis (FDA), fuzzy-c means (FCM) and vector quantization (Jiang et al., 2018; Shi et al., 2018; Tan & Xu, 2020). These techniques are employed to extract information from a dataset, which may simplify the multidimensional data to a lower approximation of the dimensions, as in case of PCA. The processed data is presented in the form of two or three principal components or factors. Similarly, different techniques have characteristic functions to process available data to generate desired reports, and they are selective for different applications; however, a combination of these algorithms may also be used for investigations (Jiang et al., 2018).

Promising applications of electronic tongues include

- To mask the taste of oral pharmaceutical formations for geriatric and pediatric patients, as unpleasant taste may discourage them from pharmacotherapy (Wasilewski et al., 2019)
- To detect adulterants and drug residues in milk which are added to present the illusion of quality nutrients or prolong the shelf life of milk and may cause severe damage to human health (Jiang et al., 2018)

- To differentiate plant oils to prevent adulteration with cheap fat sources (Gliszczyńska-Świgło & Chmielewski, 2017)
- Differentiation of honey on the basis of different botanical origins has been successfully done using an impedance spectroscopy-based electronic tongue, and the sensor was found to be very sensitive to the electrical conductivity of the honey (Elamine et al., 2019)
- A bioartificial tongue comprising chemically responsive dyes and metalloporphyrins as potential sensors has been successfully investigated for colorimetric change for protein identification. Distinct patterns were obtained for pure, mixed and denatured proteins (Hou et al., 2011)
- A significant achievement in electronic tongues has been their ability to characterize "aftertaste", which is an important parameter of a sensory profile. The residual taste of the targets on the sensors was analyzed by sending the sensors back to the reference solutions (Ross, 2021; Yin et al., 2021). Moreover, recent advances in applications of electronic tongues are presented in Table 8.2

Table 8.2   Recent Advances in Electronic Tongues for Potential Food Industry Applications

Electronic Tongue Design Specifications	Key Components Studied	Reference(s)
TS-sa402b electronic tongue (artificial lipid membrane sensors)	Evaluation of salt concentration, moisture content, taste components, pH, free amino acids and taste activity value of dry cured pork with different levels of salt	(Tian et al., 2020)
Voltametric electronic tongue (DropSens, Asturias, Spain)	Assessment of different parameters of phenolic contents in eight types of Spanish red wine and their correlation study with tristimulus color parameters	(Garcia-Hernandez et al., 2020)
Potentiometric electronic tongue (lab made, multiple sensors)	Study of influence of malaxation temperature on organoleptic characteristics and	(Marx et al., 2021)

Electronic Tongue Design Specifications	Key Components Studied	Reference(s)
	compositional parameters of olive oil; electronic tongue discriminated different samples with desired accuracy	
Potentiometric electronic tongue (lab made, two cylindrical sensor arrays)	Assessment of effect of thermal sterilization treatment on sensorial and physico-chemical characteristics of black olives; successful discrimination among different samples with electronic tongue	(Martín-Vertedor et al., 2020)
SA402B electronic tongue (five taste sensors; Insent, Tokyo, Japan)	Low-temperature long-term processing treatments for beef were successfully distinguished; effect of given treatments on taste traits and non-volatile flavor compounds was analyzed	(Ismail et al., 2020)

8.5 Biomimetic Membranes

The efficiency and complexity of biological systems has been inspiring humankind to explore such principles for technological applications. The structure–functionality relationship of biological molecules has been understood to a certain extent in the recent past, and notable efforts are being made to duplicate their performance for membrane technology. Biological membranes define the boundary between cell organelles and outer environment; however, at the same time, they show remarkable transport of selective matter in and out of living cells (Nielsen, 2009). These membranes carry out solvent and substrate transport with exceptional accuracy and selectivity in addition to brilliant anti-fouling strategies both at micro- and macro-level transport phenomena (Shen et al., 2014).

8.5.1 Biomimetic Membrane Approaches

Biological membranes have a unique set of working mechanisms which can be replicated for artificial membrane technologies to carry out a variety of functions in the food industry. Shen et al. (2014) categorized biological membranes into five paradigms: lipid bilayers, membrane proteins, surface proteins, carrier-mediated transport and anti-fouling approaches. Lipid bilayers may be referred to as canonical boundaries where biological transport of matter is being carried out by active transporters and ion channels. These membranes have been characterized as non-porous boundaries between intracellular and extracellular compartments of the cells where diffusion is the key mechanism of transport phenomenon. Shen et al. (2014) reported that membranes based on this technique are best suited for forward and reverse osmosis applications in the food industry. For the material movement in or out of the membrane, it should first enter the lipid bilayer, and then diffusion occurs within the hydrocarbon section. Another approach to developing biomimetic membranes could be simulation of activity of surface-layer proteins present in the exterior wall of prokaryotic cells (Sleytr et al., 1999). Surface-layer proteins are attached to lipid bilayers through noncovalent bonds and play the role of surface recognition, mechanical protection and ion and molecular traps. They have been found to exhibit iso-porous morphology, which makes them ideal candidates for biomimetic ultrafiltration membranes for food uses (Sára & Sleytr, 1987). In addition to potential membrane material, they have been reported to show desired anti-fouling properties; however, membrane instability and scale-up could be potential challenges (Shen et al., 2014). Further, ionophore-mediated movement of matter could be another technique for biomimetic membranes. Ionophores are charged particles, mainly of microbial origin, present in the lipid bilayer. They are composed of macrocyclic peptides with compatible ligands such as oxygen for the ability to bind to specific ions. They are potential carriers of ions across the membrane, as they can facilitate interaction with the lipid bilayer due to a hydrophobic exterior along with affinity towards ions because of their hydrophilic interior (Sengur-Tasdemir et al., 2019). Ionophores may be classified as channel formers and mobile carriers on the basis of their mode of action. Channel formers could be seen as steady structures allowing ions to pass through on the order of thousands per second; however, mobile carriers attach to ions, cross the lipid bilayer and then release the ions inside cells (Stillwell, 2016). Moreover, channel-mediated membranes could be based on block co-polymer transport or membrane protein-mediated transport. Block copolymers are amphiphilic assemblies which may be deblock or triblock with both hydrophobic and hydrophilic ends. They possess lower gas and water permeability with high chemical and mechanical strength (Purkait et al., 2018). Membrane protein-mediated transport or separation can be dedicated to functionality of proteins as promising channels, transporters and pumps (Shen et al., 2014). Membrane proteins have passive pores which may act as channels for the movement of substrate across the membrane, and the transport is driven by the concentration gradient, for instance, water channels, ion channels, porins and gap junctions (Lodish et al., 2008). Further, protein transporters use different gradients for the flow of substrate across the membrane such as concentration gradient for single solute movement (uniporters), electrochemical gradient for one solute to another in the same direction (symporters) and energy from an electrochemical gradient from one solute to another in the opposite direction (antiporters). Likewise, proteins as pumps are also based on a concentration gradient to move ions or molecules in the opposite direction using light or chemical energy sources (Shen et al., 2014).

8.5.2 Anti-Fouling Strategies for Biomimetic Membranes

- The Zwitter ionic mechanism of anti-fouling includes formation of a hydration shell by a number of water molecules through electrostatic interactions. The hydration shell prevents foulants passing through the membrane, as they must have a large amount of energy to break the shell and block the membrane (He et al., 2016)
- Release of anti-fouling compounds to prevent the adhesion of foulants on the surface of the membrane (Shen et al., 2014)
- Modification in the surface of the membranes has been reported to increase anti-fouling properties such as decreased surface roughness, increased hydrophilicity, increasing surface charge and use of biomimetic surface or thin layer surfaces (Rana & Matsuura, 2010)
- Electrostatic and hydrophobic interactions between the surface material and foulants also enhance the anti-fouling mechanism of biomimetic membranes
- Modification in the surface topography so as to obtain super-hydrophilicity or super-hydrophobicity at the surface, which could prevent interaction between foulants and the membrane surface (Purkait et al., 2018)

8.5.3 Recent Applications of Biomimetic Membranes

- Fabrication of desalination membranes has been done to retain a higher level of salt solutes and possess better chemical and mechanical stress with high water permeability (Giwa et al., 2017)
- A biomimetic artificial membrane with aquaporin proteins has been successfully tested for concentration of sugarcane juice

almost to its saturation concentration with no significant change in the nutritional composition (Akhtar et al., 2021)

- Incorporation of polymersomes in biomimetic membranes resulted in enhanced permeability of water by 30%, and use of aquaporin polymersomes further increased water permeability by 50% (Górecki et al., 2020)
- An anti-oil-fouling biomimetic membrane was developed for oil–water separation using decomposition of proanthocyanidins and aminopropyltriethoxysilane with super-hydrophilic and super-lipophilic characters. The oil rejection was found to be more than 99% without any fouling (Zhao et al., 2021)
- Biomimetic artificial water channels were developed for reverse osmosis desalination of water. The resultant membrane presented more than 99.5% rejection of salt and 91% rejection of boron with significant reduction (12%) in energy required for operation (Di Vincenzo et al., 2021)

8.6 Conclusion and Future Prospects

Nature has been addressing the problems of the environment for more than 3 billion years in a wonderful way and encouraging human beings to design novel technologies and objects. Biomimetic techniques have found huge potential in the food industry for the design of novel food products and processing technologies with better performance. Cultured meat, biomimetic membranes, electronic tongues and biomimetic milk are some of the recent successful applications of bioinspired designs. Despite the huge investment of resources and time, biomimicry in the food industry is still in its infancy. Successful models have been developed by scientists; however, their scale-up and cost still need to be optimized. More ideas from nature must be explored for product and technology innovation, as biomimetics will be the future of the food industry owing to environmentally friendly approaches and an excellent level of accuracy.

References

Akhtar, A., Singh, M., Subbiah, S., & Mohanty, K. (2021). Sugarcane juice concentration using a novel aquaporin hollow fiber forward osmosis membrane. *Food and Bioproducts Processing, 126*, 195–206.

Arshad, M. S., Javed, M., Sohaib, M., Saeed, F., Imran, A., & Amjad, Z. (2017). Tissue engineering approaches to develop cultured meat from cells: A mini review. *Cogent Food & Agriculture, 3*(1), 1320814.

Aziz, M. S. (2016). Biomimicry as an approach for bio-inspired structure with the aid of computation. *Alexandria Engineering Journal*, *55*(1), 707–714.

Bar-Cohen, Y. (2006). Biomimetics—Using nature to inspire human innovation. *Bioinspiration & Biomimetics*, *1*(1), P1.

Bhat, Z. F., & Fayaz, H. (2011). Prospectus of cultured meat—Advancing meat alternatives. *Journal of Food Science and Technology*, *48*(2), 125–140.

Bhondekar, A. P., Kaur, R., Kumar, R., Vig, R., & Kapur, P. (2011). A novel approach using dynamic social impact theory for optimization of impedance-tongue (iTongue). *Chemometrics and Intelligent Laboratory Systems*, *109*(1), 65–76.

Bhushan, B. (2009). Biomimetics: Lessons from nature–an overview. *Philosophical Transactions of the Royal Society A: Mathematical, Physical and Engineering Sciences*, *367*(1893), 1445–1486.

Bourlieu, C., Ménard, O., De La Chevasnerie, A., Sams, L., Rousseau, F., Madec, M.-N., Robert, B., Deglaire, A., Pezennec, S., & Bouhallab, S. (2015). The structure of infant formulas impacts their lipolysis, proteolysis and disintegration during *in vitro* gastric digestion. *Food Chemistry*, *182*, 224–235.

Bourlieu, C., Paboeuf, G., Chever, S., Pezennec, S., Cavalier, J.-F., Guyomarc'h, F., Deglaire, A., Bouhallab, S., Dupont, D., & Carrière, F. (2016). Adsorption of gastric lipase onto multicomponent model lipid monolayers with phase separation. *Colloids and Surfaces B: Biointerfaces*, *143*, 97–106.

Bourlieu-Lacanal, C., Deglaire, A., de Oliveira, S. C., Ménard, O., Le Gouar, Y., Carrière, F., & Dupont, D. (2017). Towards infant formula biomimetic of human milk structure and digestive behaviour. *OCL Oilseeds and Fats Crops and Lipids*, *24*(2), D206.

Chen, M., & Sagis, L. M. (2019). The influence of protein/phospholipid ratio on the physicochemical and interfacial properties of biomimetic milk fat globules. *Food Hydrocolloids*, *97*, 105179.

Ciosek, P., & Wróblewski, W. (2007). Sensor arrays for liquid sensing–electronic tongue systems. *Analyst*, *132*(10), 963–978.

Collins, C. A., Zammit, P. S., Ruiz, A. P., Morgan, J. E., & Partridge, T. A. (2007). A population of myogenic stem cells that survives skeletal muscle aging. *Stem Cells*, *25*(4), 885–894.

Conway, V., Gauthier, S. F., & Pouliot, Y. (2014). Buttermilk: Much more than a source of milk phospholipids. *Animal Frontiers*, *4*(2), 44–51.

Di Vincenzo, M., Tiraferri, A., Musteata, V.-E., Chisca, S., Sougrat, R., Huang, L.-B., Nunes, S. P., & Barboiu, M. (2021). Biomimetic artificial water channel membranes for enhanced desalination. *Nature Nanotechnology*, *16*(2), 190–196.

Dong, F., Padua, G. W., & Wang, Y. (2013). Controlled formation of hydrophobic surfaces by self-assembly of an amphiphilic natural protein from aqueous solutions. *Soft Matter*, *9*(25), 5933–5941.

Dong, F., Zhang, M., Tang, W.-W., & Wang, Y. (2015). Formation and mechanism of superhydrophobic/hydrophobic surfaces made from amphiphiles through droplet-mediated evaporation-induced self-assembly. *The Journal of Physical Chemistry B, 119*(16), 5321–5327.

Elamine, Y., Inácio, P. M., Lyoussi, B., Anjos, O., Estevinho, L. M., da Graça Miguel, M., & Gomes, H. L. (2019). Insight into the sensing mechanism of an impedance based electronic tongue for honey botanic origin discrimination. *Sensors and Actuators B: Chemical, 285*, 24–33.

Gallier, S., Gragson, D., Jiménez-Flores, R., & Everett, D. (2010). Using confocal laser scanning microscopy to probe the milk fat globule membrane and associated proteins. *Journal of Agricultural and Food Chemistry, 58*(7), 4250–4257.

Garcia-Hernandez, C., Salvo-Comino, C., Martin-Pedrosa, F., Garcia-Cabezon, C., & Rodriguez-Mendez, M. L. (2020). Analysis of red wines using an electronic tongue and infrared spectroscopy. Correlations with phenolic content and color parameters. *LWT, 118*, 108785.

Giwa, A., Hasan, S. W., Yousuf, A., Chakraborty, S., Johnson, D. J., & Hilal, N. (2017). Biomimetic membranes: A critical review of recent progress. *Desalination, 420*, 403–424.

Glatz, R., & Bailey-Hill, K. (2011). Mimicking nature's noses: From receptor deorphaning to olfactory biosensing. *Progress in Neurobiology, 93*(2), 270–296.

Gliszczyńska-Świg\lo, A., & Chmielewski, J. (2017). Electronic nose as a tool for monitoring the authenticity of food. A review. *Food Analytical Methods, 10*(6), 1800–1816.

Górecki, R., Reurink, D. M., Khan, M. M., Sanahuja-Embuena, V., Trzaskuś, K., & Hélix-Nielsen, C. (2020). Improved reverse osmosis thin film composite biomimetic membranes by incorporation of polymersomes. *Journal of Membrane Science, 593*, 117392.

He, M., Gao, K., Zhou, L., Jiao, Z., Wu, M., Cao, J., You, X., Cai, Z., Su, Y., & Jiang, Z. (2016). Zwitterionic materials for antifouling membrane surface construction. *Acta Biomaterialia, 40*, 142–152.

Hernell, O., Timby, N., Domellöf, M., & Lönnerdal, B. (2016). Clinical benefits of milk fat globule membranes for infants and children. *The Journal of Pediatrics, 173*, S60–S65.

Hou, C., Dong, J., Zhang, G., Lei, Y., Yang, M., Zhang, Y., Liu, Z., Zhang, S., & Huo, D. (2011). Colorimetric artificial tongue for protein identification. *Biosensors and Bioelectronics, 26*(10), 3981–3986.

Ismail, I., Hwang, Y.-H., & Joo, S.-T. (2020). Low-temperature and long-time heating regimes on non-volatile compound and taste traits of beef assessed by the electronic tongue system. *Food Chemistry, 320*, 126656.

Ivarsson, P., Krantz-Rülcker, C., Winquist, F., & Lundström, I. (2005). A voltammetric electronic tongue. *Chemical Senses, 30*(suppl_1), i258–i259.

Jiang, H., Zhang, M., Bhandari, B., & Adhikari, B. (2018). Application of electronic tongue for fresh foods quality evaluation: A review. *Food Reviews International, 34*(8), 746–769.

Kallio, M. J., Salmenperä, L., Siimes, M. A., Perheentupa, J., & Miettinen, T. A. (1992). Exclusive breast-feeding and weaning: Effect on serum cholesterol and lipoprotein concentrations in infants during the first year of life. *Pediatrics, 89*(4), 663–666.

Kanatous, S. B., & Mammen, P. P. (2010). Regulation of myoglobin expression. *Journal of Experimental Biology, 213*(16), 2741–2747.

Lepora, N. F., Verschure, P., & Prescott, T. J. (2013). The state of the art in biomimetics. *Bioinspiration & Biomimetics, 8*(1), 013001.

Lodish, H., Berk, A., Kaiser, C. A., Kaiser, C., Krieger, M., Scott, M. P., Bretscher, A., Ploegh, H., & Matsudaira, P. (2008). *Molecular cell biology*. Macmillan.

Lopez, C., Cauty, C., & Guyomarc'h, F. (2015). Organization of lipids in milks, infant milk formulas and various dairy products: Role of technological processes and potential impacts. *Dairy Science & Technology, 95*(6), 863–893.

Lopez, C., & Ménard, O. (2011). Human milk fat globules: Polar lipid composition and in situ structural investigations revealing the heterogeneous distribution of proteins and the lateral segregation of sphingomyelin in the biological membrane. *Colloids and Surfaces B: Biointerfaces, 83*(1), 29–41.

Martin, J.-C., Bougnoux, P., Antoine, J.-M., Lanson, M., & Couet, C. (1993). Triacylglycerol structure of human colostrum and mature milk. *Lipids, 28*(7), 637–643.

Martín-Vertedor, D., Rodrigues, N., Marx, Í. M., Veloso, A. C., Peres, A. M., & Pereira, J. A. (2020). Impact of thermal sterilization on the physicochemical-sensory characteristics of Californian-style black olives and its assessment using an electronic tongue. *Food Control, 117*, 107369.

Marx, Í. M., Rodrigues, N., Veloso, A. C., Casal, S., Pereira, J. A., & Peres, A. M. (2021). Effect of malaxation temperature on the physicochemical and sensory quality of cv. Cobrançosa olive oil and its evaluation using an electronic tongue. *LWT, 137*, 110426.

Matindoust, S., Baghaei-Nejad, M., Abadi, M. H. S., Zou, Z., & Zheng, L.-R. (2016). Food quality and safety monitoring using gas sensor array in intelligent packaging. *Sensor Review, 36*(2), 169–183.

Mauro, A. (1961). Satellite cell of skeletal muscle fibers. *The Journal of Cell Biology, 9*(2), 493–495.

Misra, A. K. (1982). Infant feeding: Breast or bottle? *Dairy Guide, 2*, 22–26.

Nielsen, C. H. (2009). Biomimetic membranes for sensor and separation applications. *Analytical and Bioanalytical Chemistry, 395*(3), 697–718.

Noor, S., Radhakrishnan, N. S., & Hussain, K. (2016). Newer trends and techniques adopted for manufacturing of *in vitro* meat through "tissue-engineering" technology: A review. *International Journal of Biotech Trends and Technology, 19,* 14–19.

Oosting, A., Engels, E., Kegler, D., Abrahamse, M., Teller, I., & Van Der Beek, E. (2011). A more breast milk-like infant formula reduces excessive body fat accumulation in adult mice. *Pediatric Research, 70*(5), 837–837.

Orlandi, G., Calvini, R., Foca, G., Pigani, L., Simone, G. V., & Ulrici, A. (2019). Data fusion of electronic eye and electronic tongue signals to monitor grape ripening. *Talanta, 195,* 181–189.

Post, M. J. (2012). Cultured meat from stem cells: Challenges and prospects. *Meat Science, 92*(3), 297–301.

Purkait, M. K., Sinha, M. K., Mondal, P., & Singh, R. (2018). Biologically responsive membranes. In *Interface science and technology* (Vol. 25, pp. 145–171). Elsevier.

Rana, D., & Matsuura, T. (2010). Surface modifications for antifouling membranes. *Chemical Reviews, 110*(4), 2448–2471.

Ross, C. F. (2021). Considerations of the use of the electronic tongue in sensory science. *Current Opinion in Food Science, 40,* 87–93.

Sánchez-Hidalgo, V. M., Flores-Huerta, S., Matute-González, G., Urquieta-Aguila, B., Bernabe-García, M., & Cisneros-Silva, I. E. (2000). A fortifier comprising protein, vitamins, and calcium-glycerophosphate for preterm human milk. *Archives of Medical Research, 31*(6), 564–570.

Sára, M., & Sleytr, U. B. (1987). Production and characteristics of ultrafiltration membranes with uniform pores from two-dimensional arrays of proteins. *Journal of Membrane Science, 33*(1), 27–49.

Sarkar, S. (2003). Recent innovations in cultured milk products for infants. *Nutrition & Food Science, 33*(6), 268–272.

Sengur-Tasdemir, R., Tutuncu, H. E., Gul-Karaguler, N., Ates-Genceli, E., & Koyuncu, I. (2019). Biomimetic membranes as an emerging water filtration technology. In *Biomimetic lipid membranes: Fundamentals, applications, and commercialization* (pp. 249–283). Springer.

Shen, Y., Saboe, P. O., Sines, I. T., Erbakan, M., & Kumar, M. (2014). Biomimetic membranes: A review. *Journal of Membrane Science, 454,* 359–381.

Shi, H., Zhang, M., & Adhikari, B. (2018). Advances of electronic nose and its application in fresh foods: A review. *Critical Reviews in Food Science and Nutrition, 58*(16), 2700–2710.

Slack, J. M. W. (2008). Origin of stem cells in organogenesis. *Science, 322*(5907), 1498–1501.

Sleytr, U. B., Messner, P., Pum, D., & Sára, M. (1999). Crystalline bacterial cell surface layers (S layers): From supramolecular cell structure to biomimetics and nanotechnology. *Angewandte Chemie International Edition, 38*(8), 1034–1054.

Smith, M. M., & Kuhn, L. (2000). Exclusive breast-feeding: Does it have the potential to reduce breast-feeding transmission of HIV-1? *Nutrition Reviews, 58*(11), 333–340.

Stillwell, W. (2016). *An introduction to biological membranes: Composition, structure and function.* Elsevier.

Tan, J., & Xu, J. (2020). Applications of electronic nose (e-nose) and electronic tongue (e-tongue) in food quality-related properties determination: A review. *Artificial Intelligence in Agriculture, 4,* 104–115.

Tian, X., Li, Z. J., Chao, Y. Z., Wu, Z. Q., Zhou, M. X., Xiao, S. T., Zeng, J., & Zhe, J. (2020). Evaluation by electronic tongue and headspace-GC-IMS analyses of the flavor compounds in dry-cured pork with different salt content. *Food Research International, 137,* 109456.

Wasilewski, T., Migoń, D., Gębicki, J., & Kamysz, W. (2019). Critical review of electronic nose and tongue instruments prospects in pharmaceutical analysis. *Analytica Chimica Acta, 1077,* 14–29.

Winquist, F., Krantz-Rülcker, C., & Lundström, I. (2002). Electronic tongues and combinations of artificial senses. *Sensors Update, 11*(1), 279–306.

Yin, X., Lv, Y., Wen, R., Wang, Y., Chen, Q., & Kong, B. (2021). Characterization of selected Harbin red sausages on the basis of their flavour profiles using HS-SPME-GC/MS combined with electronic nose and electronic tongue. *Meat Science, 172,* 108345.

Zandonella, C. (2003). Tissue engineering: The beat goes on. *Nature, 421*(6926), 884–887.

Zeng, J., Yu, W., Dong, X., Zhao, S., Wang, Z., Liu, Y., Wong, M.-S., & Wang, Y. (2019). A nanoencapsulation suspension biomimetic of milk structure for enhanced maternal and fetal absorptions of DHA to improve early brain development. *Nanomedicine: Nanotechnology, Biology and Medicine, 15*(1), 119–128.

Zhang, G., Zhao, X., Li, X., Du, G., Zhou, J., & Chen, J. (2020). Challenges and possibilities for bio-manufacturing cultured meat. *Trends in Food Science & Technology, 97,* 443–450.

Zhao, Y., Yang, X., Yan, L., Bai, Y., Li, S., Sorokin, P., & Shao, L. (2021). Biomimetic nanoparticle-engineered superwettable membranes for efficient oil/water separation. *Journal of Membrane Science, 618,* 118525.

Zou, X.-Q., Guo, Z., Huang, J.-H., Jin, Q.-Z., Cheong, L.-Z., Wang, X.-G., & Xu, X.-B. (2012). Human milk fat globules from different stages of lactation: A lipid composition analysis and microstructure characterization. *Journal of Agricultural and Food Chemistry, 60*(29), 7158–7167.

Nanofluids in Thermal Processing of Foods

Mehrajfatema Zafar Mulla, Reshma Bhatnagar, and Aakriti Kapoor

9.1 Introduction

Thermal processes are ubiquitous in the food industry, and the safety and quality of food are intricately linked to thermal processing, for example, pasteurization, sterilization, ultra–high temperature processing, refrigeration, boiling, frying and baking. These processes not only preserve the food but also make the food ingestible and more flavourful and improve its texture. While these advances have given rise to convenience food, industrially processed food for most consumers has become synonymous with junk and nutrition-poor food, and to some extent for good reason. While thermal processing of foods has made packaged food safe to eat over a long period of time, a long duration of heating at high temperature has the tendency to denature and destroy key nutrients in foods. Preserving the nutrient content whilst maintaining long-term sterility is therefore an area of continuing research. It is posited that enhanced thermal efficiency can reduce processing time, thus reducing the breakdown of thermo-labile components of food. It is thus hoped that with their greater thermal efficiency, nanofluids may be able to help preserve nutrients better by bringing down the overall processing time whilst still creating conditions suitable for microbial control (Jabbari et al., 2018).

DOI: 10.1201/9781003217138-9

Indeed, thermal processing is a key energy expense, with food processing a major consumer of energy (Alia et al., 2019). Equipment in which thermal processing is undertaken is therefore optimized considerably to minimize heat loss and ensure energy efficiency. Heat transfer media used in vessels to control temperature need to be chosen optimally. Several selection criteria such as food composition, equipment type and heat transfer fluid are responsible for optimum results. Further, these selection criteria include knowledge of the process, such as food composition, temperature requirements and differentials; knowledge of the vessel/equipment, such as composition of the vessel, the size and piping; and knowledge of prospective heat transfer fluid: specific heat capacity, melting and boiling points, viscosity, oxidative stability, and corrosion characteristics.

To reconcile the world's increasing energy needs with the goals of cleaner and greener energy, more energy-efficient systems are the way forward. The use of nanofluids as heat transfer fluids is one of the options to achieve this goal.

9.2 Heat Transfer Fluids

Heat transfer occurs by the processes of conduction, convection and radiation. Fluids, liquids and gases that transfer heat from one point to another are called heat transfer fluids. These fluids can be used as media to heat or cool the system of interest. Heating and cooling coils and plate heat exchangers are typical examples of systems requiring heat transfer fluids.

Conventional heat transfer fluids often contain water, oil or glycol as their primary constituent. Other popular heat transfer fluids are based on mineral oil compositions. Heat transfer fluids for refrigeration are often glycol-based systems, as it provides considerable fluidity at low temperatures (Banisharif et al., 2020). These fluids are often combined with water to bring down cost while at the same time providing an adequate drop in freezing point (Banisharif et al., 2020). Certainly, newer technologies based on esters or silicones have further improved key characteristics of heat transfer fluids extensively (Perrier and Beroual, 2010). Nanofluids are another promising novel technology that may dramatically enhance heat transfer efficiency.

The most significant factor that affects heat transfer is thermal conductivity, and the following variables have an influence on thermal conductivity: particle size, density and free electrons. The convective heat transfer coefficient influences how quickly heat is transferred; therefore, choosing the right heat transfer fluid requires consideration of these aspects. These factors are viscosity, velocity, heat flux, surface morphology and flow characteristics of fluid. In such fluids, combining metal-based nanoparticles

(NPs) can increase the heat transfer coefficient and subsequently the heat transfer rate.

9.2.1 Composition and Preparation of Nanofluids

Nanofluids consist of the following two components at the very least:

1. A base fluid, which facilitates the transfer of heat by movement
2. NPs, which improve the base fluid's capacity for heat transmission and range in size from 1 to 100 nm

To enhance the functionality of the nanofluid, it can additionally include extra stabilizers (often surfactants) or other chemicals.

The promising effect of nanofluids is based on the fact that metals have higher thermal conductivity in solid form than fluid form. Therefore, increased thermal conductivity rates can be expected in fluids containing suspended solid particles in comparison to base fluids like water and ethylene glycol. An inclusive array of NPs has been employed for the production of nanofluids; some of those commonly used are as follows:

1. Carbon NPs such as single- or multi-walled carbon nanotubes, graphene, graphene oxide (GO), diamond, fullerene and graphite.
2. Metal NPs, namely silver, aluminium, gold, copper and iron.
3. Metal oxide NPs such as aluminium oxide (Al_2O_3), ceric oxide (CeO_2), copper oxide (CuO), iron oxide (Fe_3O_4), titanium dioxide (TiO_2) and zinc oxide (ZnO).
4. Others such as silicon, aluminium nitride-carbon (AlN-C), carbon monoxide (CO), cobalt ferrite ($CoFe_2O_4$), silicon carbide (SiC), Field's alloy NPs, zinc bromide ($ZnBr_2$) and silicon dioxide (SiO_2).

The first and most important stage in using NPs as a heat transfer fluid is the fabrication of nanofluids. The preparation of nanofluids is typically done in a one- or two-step process. For the one-step process, NPs are fabricated in situ, while for the two-step process, NPs are first constructed using conventional methods and are further suspended in a base fluid (Ghadimi et al., 2011). While it has been proposed that a one-step procedure for producing nanofluids can give them more stability than a two-step process does, the complexity of one-step processes and the lack being able to select from a range of NPs likely nullifies most advantages of nanofluid production using one-step processes, especially since nanofluids prepared by two-step processes can be stabilized using appropriate additives and surfactants

(Kaggwa and Carson, 2019). Due to the associated costs, the physics-based technique limits the large-scale manufacture of nanofluids, whereas the chemical-based approach introduces a significant amount of contaminants during processing (Yu and Xie, 2012). In order to reduce the agglomeration of NPs, low-vapour-pressure fluids can only be administered in a single phase (Xian et al., 2018).

The key advantages of using the two-step process are the variety of NPs that can be utilized and the lack of complexity in creating nanofluid suspensions. The two-step method is shown in Figure 9.1. In this method, nanofluids are synthesized either on laboratory scale or industrially, followed by mechanical stirring in the base fluid for the required time per the type, shape and size of NPs.

Commonly utilized based fluids include water, ethylene glycol and oil, which are also traditional heat transfer fluids (Okonkwo et al., 2021). While the mechanism of action of heat transfer has not yet been fully understood (several studies describe enhanced heat transfer properties as "anomalous"; Choi et al., 2001), it is attributed to the increased surface area of nanoparticles (Simpson et al., 2019). Furthermore, compared to conventional heat transfer fluids or macrofluids, nanofluids offer less flow channel obstruction and reduced friction (Jama et al., 2016). A comparison of thermal conductivity data for base fluids and nanofluids is given in Table 9.1.

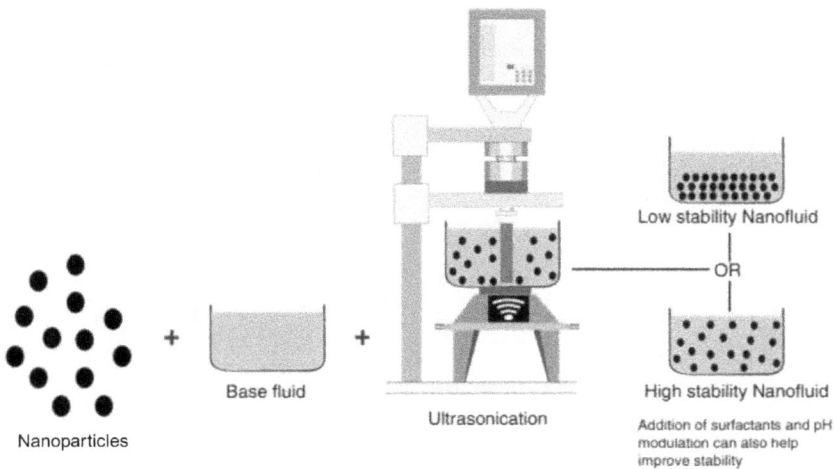

Figure 9.1 Two-step method for the preparation of nanofluids.

Source: Okonkwo et al. (2020)

Table 9.1 Thermo-Physical Properties of Nanofluids Used for Food Processing

Food Sample	Nanofluid composition	NP Diameter (nm)	NP Concentration (%)	Density (kg/m3)	Viscosity (mPa/s)	Specific Heat Capacity (J/kg K)	Thermal Conductivity (W/m.K)	Instrument Type/Thermal Processing system	Reference
Tomato juice	Al_2O_3–water	20	0	996	0.611	4181	0.610	Shell and tube heat exchanger	Jafari et al. (2017).
			1	1025	0.627	4056	0.628		
			2	1054	0.645	3937	0.645		
			3	1083	0.664	3825	0.664		
			4	1112	0.682	3719	0.682		
Milk	Al_2O_3–water	50 nm	0.3% of its volume concentration	1016.352	0.803 77	4170	1.0085	Plate heat exchanger	Tamilselvan et al. (2017).
Not given	Graphene sheets-water	–	0.050 wt%	1023.8	0.9429	4009.4	0.704	Shell and tube heat exchanger	Ghozatloo et al. (2014).
			0.075 wt%	1038.4	0.9698	3924.9	0.791		
			0.1 wt%	1053.5	0.9977	3840.1	0.689		
Not given 60:40 EG/W				1080.00	3.99	3157.19	0.360	compact heat exchangers	Ray et al. (2014).
	Al2O3 EG/W	45	1	1105.23	4.4799	3079.04	0.400		
	CuO EG/W	–	1	1134.25	4.63	3006.39	0.402		
	SiO2 EG/W	–	1	1091.22	4.38	3108.42	0.379		

9.2.2 Nanoparticle Preparation

The preparation of NPs can be undertaken in a variety of ways, physical or chemical, and can utilize bottom-up or top-down approaches. After exploring all the available methods for NP preparation, a few common methods are listed here.

1. Chemical vapour deposition (Park et al., 2014)
2. Sol-gel process (Singh et al., 2014)
3. Grinding/shearing/ball mill (Seong et al., 2018)
4. Hummers' method (Hummers and Offeman, 1958)

The selection of the preparation method is more or less responsible for the resultant morphology of NPs and their behaviour (Dhand et al., 2015). Chen et al. (2009) successfully altered the morphology of Fe_3O_4 to form nanowires, cubes, biscuit-shaped or spherical NPs by altering the reaction conditions, inclusion of organic solvents and quantity of surfactants used.

9.2.3 Challenges in Manufacture and Stabilization

A wide variety of NP substances can be used to make nanofluids. Common NP materials include metal oxides, ceramics, sulphides, alloys and carbon-based varieties (Khan et al., 2017). Hybrid nanofluids consisting of more than one particle material may also be formulated to provide even greater heat transfer enhancements via synergistic effects (Sarkar et al., 2015). Different sizes and shapes of NPs of the same material also impact the thermal properties of base fluids. Cube-shaped NPs, for example, have been shown to have superior thermal conductivity when compared with spherical or rod-shaped NPs (Maheshwary et al., 2017).

The biggest obstacle preventing nanofluids from being used extensively in industry is the inability to maintain their stability. The tendency of NPs to form aggregates due to immense van der Waals forces make these formulations inherently unstable, requiring strategies to ensure homogenization and balance the attractive forces between particles (Ghadimi et al., 2011). Stable nanofluids thus require further processing or the addition of stabilizers for any potential useful formulation.

A commonly used strategy is to formulate stable nanofluids is the addition of surfactants like sodium dodecyl sulphate, oleic acid or gum arabic that impact the interfacial interaction properties of particles and base fluids (Yu et al., 2012, Tripathy et al., 2016).

Surfactants/surface-active agents often contain hydrophobic and hydrophilic moieties in the same molecule, thereby modulating repulsive forces between NPs and base fluids. The hydrophilic-lipophilic balance (HLB) of the surfactant molecule is a measure of water or oil affinity that can be used for the selection of the appropriate surfactant for the base fluids.

Selection of an appropriate surfactant can also depend on surfactant charge; that is, is it cationic, anionic, amphoteric (zwitter ionic surfactants whose charge depends on pH) or non-ionic (carrying no charge)? The general rule of thumb while selecting a surfactant for dispersing NPs is that water-soluble surfactants should be used for polar base fluids and oil-soluble surfactants for non-polar base fluids (Ghadimi et al., 2011).

Another common method to stabilize nanofluids is ultrasonication, which breaks down agglomerations/aggregates and homogenizes the suspensions. Homogenization/high-speed stirrers have also been used, though occasional problems at the time of processing have been noted (Ghadimi et al., 2011).

Other less-studied methods to stabilize nanofluid solutions include functionalization of NPs (for example, by adding an appropriate functional group such as –COOH) to improve dispersibility (Esfe et al., 2014) or pH control.

9.3 Methods to Evaluate Nanofluid Stability

9.3.1 UV-Vis Spectrophotometry

The addition of NPs in base fluids has a tendency to change the optical characteristics of the base fluid. These changes in optical properties relate to particle size, providing an indication of the degree of agglomeration and thus the stability of the suspension (Yu and Xie, 2012). Experimentally the linear relationship between concentration and absorbance as described by Beer-Lambert's law correlates well for nanofluids, making it a popular method to determine NP concentration as well (Sharif et al., 2017).

9.3.2 Zeta Potential

The electric forces between like and unlike particles determine the tendency of agglomeration of particles. Zeta potential is the electric charge that is generated at the interface of the particle in suspension and the base fluid surrounding it and thus provides an idea of whether there will be a tendency of the NPs to agglomerate. Zeta potential is influenced by a number of variables, including pH, ionic strength, additive concentration and temperature (Lu and Gao, 2010).

The determination of zeta potential can be done using a pair of electrodes, wherein the particle velocity determined by Doppler techniques is measured as a function of voltage (Selvamani, 2019). This is the electrophoretic light scattering method. A dynamic light scattering method may also be used for the determination of zeta potential, wherein a laser beam provides a monochromatic light source, which scatters when it comes in contact with NPs in the sample. Photodiodes capture the backscattering (Bhattacharjee, 2016).

Knowledge of the zeta potential is possibly the most valuable way to predict the stability of a nanofluid. Accurately predicting or modelling zeta potential can therefore be an invaluable tool for nanofluid formulators and continues to be an active area of research. A computational model to determine zeta potential with good fitting with experimental data was suggested by Biriukov et al. (2020). Mikolajczyk et al. (2015) developed the quantitative structure-property (QSPR) model to predict zeta potential, which can be used for in silico determination.

9.3.3 Other Methods

While spectrometry and zeta potential measurements are the most popular methods of determining stability of nanofluids, many other methods may be used. The relatively simple method of sediment photograph capturing involves photographing the sedimentation of nanofluids over time. Microscopic techniques such as transmission electron microscopy (TEM) and scanning electron microscopy (SEM) allow for the actual visualization of the degree of aggregations (Ghadimi et al., 2011). Light-scattering techniques have been used for over a century to measure dispersion of colloidal suspensions (Tyndall, 1868; Carvalho et al., 2018).

9.4 Factors Affecting Nanofluids and Predicting Nanofluid Behaviour

9.4.1 Thermal Conductivity

One of the most important qualities relating to heat transport is thermal conductivity, which has undergone substantial research (Kaggwa and Carson, 2019). The capacity of a material to transport heat is known as its thermal conductivity, which measures the amount of thermal energy that is transferred over a temperature gradient in a given amount of time. The material's inherent quality of thermal conductivity is represented by the following equation:

$$K = dQ / A\Delta T \qquad (9.1)$$

The transient hot wire method (THW), which measures the rise in temperature in a linear heat source (the wire) when introduced in the nanofluid, can be used to experimentally determine it. (Merckx et al., 2012).

There have been a number of models proposed to calculate the thermal conductivity of suspensions, most of which are based on Maxwell's groundbreaking research on the issue. Maxwell predicted the thermal conductivity on the volume fractions of spherical particles in base fluids:

$$K_{nf} = \left[(K_p + 2K_{bf} + 2\phi(K_p - K_{bf})) \right].K_{bf}$$
$$(K_p + 2K_{bf} - \phi(K_p - K_{bf})) \qquad (9.2)$$

The aforementioned model, however, is limited to highly diluted suspensions and overlooks environmental factors, including base fluid; pH; temperature; and particle size, shape and type (Kaggwa and Carson, 2019). Hamilton and Crosser (1962) therefore proposed a newer model, which corrected for particle shape:

$$K_{nf} = K_{bf}.K_p + (n-1).K_{bf} - (n-1).\phi(K_{bf} - K_p)$$
$$K_p + (n-1)K_{bf} + \phi(Kbf - Kp) \qquad (9.3)$$

It has been shown that the rise in thermal conductivity of nanofluids is boosted above predicted levels with regard to the fraction of NPs introduced in the base fluid.

Many models, often modifications of the previous ones to include ignored particles and environmental parameters, have been proposed. Kumar et al. (2015) reviewed over 40 models of thermal conductivity, most derivatives of the previous classical models, which try to incorporate factors that can model for anomalous heat conductivity enhancements. Computational and statistical methods such as the Monte Carlo method have also been used to model thermal conductivity (Ghadimi et al., 2011).

The impacts of Brownian motion, the creation of interfacial layers and particle clustering, particle size, morphology, temperature and surfactant choice are all variables that affect thermal conductivity, according to newer models (Kaggwa and Carson, 2019). Models also suggest that thermal conductivity is not significantly impacted at different doses of surfactant or base fluid concentration, indicating that thermal conductivity is being impacted by only the addition of NPs (Ghadimi et al., 2011).

While newer models may give more accurate results, classical models can still provide a fair estimate and serve formulators due to their simplicity. For example, Utomo et al. (2012) reported experimental thermal conductivity for alumina and titania nanofluids to be within ±5% of the values predicted by Maxwell's model. Even as several new models have shown very good correlation with selected experimental results, no model has been found to give satisfactorily accurate results routinely, and this remains an active area of research.

9.4.2 Viscosity

Viscosity is defined as the resistance to flow. Viscosity is a critical parameter in heat exchange systems, as it determines the ease of flow and the force required to move the liquid within a heat exchanger. Predicting the viscosity and flow behaviour of heat exchange fluids can therefore be helpful to nanofluid formulators.

Although it was first believed that the addition of NPs would not affect viscosity due to their size and concentration (Choi and Eastman, 1995), investigations have revealed that the addition of NPs can affect base fluid viscosity (Kaggwa and Carson, 2019).

One of the models to predict viscosity is Einstein's model. In this model, viscosity determination is based on the volume fraction of the suspended particles as well as the viscosity of the base fluid. The drawback of this method is that the model works only for very small volume fractions of idealised spherical particles:

$$\mu_{nf} = (1 + 2.5\phi).\mu_{bf} \qquad (9.4)$$

Krieger and Dougherty suggested corrections to Einstein's model, by considering the full range of particle volume fraction:

$$\mu_{nf} = (1 - \phi / \phi_m)^{-\eta\phi m}.\mu_{bf} \qquad (9.5)$$

Newer methods to predict viscosity include correlational models and artificial neural networks (ANNs) (Okonkwo et al., 2021).

9.4.3 Specific Heat

The amount of heat needed to raise a substance's temperature by 1°C per unit mass is known as specific heat. It is a measure of the heat-carrying capacity of a substance and plays a role in the heat storage ability of thermal fluids (Carrillo-Berdugo et al., 2020). Higher heat capacities allow for reduced temperature variation during heat transfer compared to substances with lower heat capacities.

Specific heats for nanofluids have not been studied as extensively as thermal conductivity, but correlation models have been developed, as have ANN approaches and machine learning models, showing excellent accuracy (Okonkwo et al., 2021).

Differential scanning calorimetry has been identified as a powerful tool to measure the specific heat of nanofluids (O'Hanley, 2012; Sadeghinezhad et al., 2016).

9.5 Modelling of Heat Transfer in Nanofluids

Several models for thermal transfer have been suggested, falling into two groups: homogenous flow models that underpredict the heat transfer co-efficient of nanofluids and dispersion models that attribute heat transfer enhancement to higher thermal conductivity and dispersion of NPs (Buongiorno, 2006). Homogenous flow models assume that conventional fluid equations may be extended for nanofluids as well; hence traditional heat transfer correlations can be used but often conflict with experimental observations.

Buongiorno (2006) attempted to correct these shortcomings by modelling the transport mechanisms of nanofluids, taking into consideration the variables that affect the nanofluid velocity, like the inertia of particles, Brownian diffusion, thermophoresis (forces on particles due to a temperature gradient), diffusiophoresis (forces on particles due to a concentration gradient), the Magnus effect (forces perpendicular to the direction of flow due to shear stress), fluid drainage and gravity, and found that thermophoresis and Brownian diffusion were the only factors to have any significant impact and proposed a new heat transfer correlation which matched well with experimental data:

$$Nu_b = 0.021.Re_b^{0.8} Pr_b^{0.5} \qquad (9.6)$$

$$Nu_b = 0.0059.[1 + 7.6286 \; \phi \left(Re_b Pr_b.d_p \; / \; D \right)^{0.001}] \; .Re_b^{0.9238} Pr_b \qquad (9.7)$$

Heat transfer by nanofluids, as discussed earlier, is thus dependent on several factors, including NP type, size, concentration, the base fluid, surfactant used, method of processing and temperature.

9.6 Application of Nanofluids in Heat Exchangers for Food Industry

Nanofluids have been suggested for use in applications for cooling and heating. Interest has been generated in the use of nanofluids as thermal fluids as well as refrigerants (Okonkwo et al., 2021; Kaggwa and Carson, 2019).

A device that transmits heat from one medium to another is a heat exchanger (Curd et al., 2001). This transfer generally occurs without the two media coming in direct contact with each other. Active heat exchangers (unlike heat sinks) use a fluid as a medium to carry heat within their systems. The medium used to transfer heat is called a heat transfer fluid, and industrially, the base fluids used in nanofluid preparations have been used for this purpose.

Recent studies have explored the use of nanofluids in a variety of heat exchangers, attempting to assess the possibility of using them in industry. The applicability of nanofluids for the plate heat exchanger, circular tube heat exchanger, double tube heat exchanger and shell and tube heat exchanger types of heat exchangers has been investigated. Table 9.1 lists the thermophysical characteristics of nanofluids used in food preparation.

Recent years have seen a thorough evaluation of the use of nanofluids as a heat transfer medium in heat exchangers for enhanced thermal conductivity and viscosity of the base fluid to expedite pressure drop.

Heat exchangers are among the essential thermal processing units in the food industry. Although food businesses have been using heat exchangers for more than a century, there is always space for improvement for greater efficiency. Heat exchangers are equipment employed for the heat transfer mechanism between different fluids. Low heat transfer rates of base fluids like ethylene glycol operated in heat exchangers provide continued motivation for improvement. Markedly increased energy requirements and concern regarding depletion of natural resources in recent years have further incentivized the industry. Traditionally water is used as a processing fluid in these heat exchangers, which takes a long time for processing, and several studies have been carried out to improve heat exchanger efficiency rates. Recently researchers observed that the use of nanofluids instead of water can reduce the processing time and eventually reduce energy consumption and improve production output. Recently Jaferi et al. (2017) reported an increase in thermal conductivity and heat transfer coefficient by 12.8 and 13%, respectively, for watermelon juice. Moreover, tomato juice processing using nanofluids at 4% concentration also showed an enhancement of thermal conductivity to 12.1% (Jafari et al., 2017).

Further chilling is also one of the important processing steps in many liquid food-processing plants, especially milk processing units.

Recently Bhattad et al. (2019) studied brine-based nanofluids containing silver, alumina and magnesia NPs in plate heat exchangers for milk chilling. An enhancement in convective and overall heat transfer coefficients, heat transfer rate, pumping power, pressure drop, entropy generation rate, irreversibility and milk flow rate was observed using a hybrid nanofluid. The maximum convective heat transfer (9.4%) and heat transfer rate (1.6%) were observed for alumina-silver-ethylene glycol nanofluids. Therefore, low temperature–based processes like chilling of milk could be easily achieved with combination hybrid nanofluid (brine-based) and heat exchangers.

9.6.1 Shell and Tube Heat Exchangers

A shell and tube type of heat exchanger allows a maximum area of contact as compared to other heat exchangers, which is an added advantage in thermal processing. It consists of an inner tube bundle surrounded by an outer tube, often known as the shell. A cross-sectional view of this type of heat exchanger is shown in Figure 9.2. Shell diameter and the sum of tubes placed inside the shell have a significant influence on the turbulence and heat transmission rate; therefore, the shell diameter should be designed to give a close fit to the bundle of tubes and maximum numbers of tubes installed in it. Apart from the number of tubes, the thickness of the tubes must be optimal to withstand internal pressure and corrosion. Installation of baffles helps to elevate the fluid flow inside the tube bundles and encompasses a prominent elevation in the heat transmission rates. Further, to achieve higher heat transfer coefficients and pressure drop, baffle spacing (distance among adjacent baffles) should be kept small (Silaipillayarputhur and Khurshid, 2019). The hot base

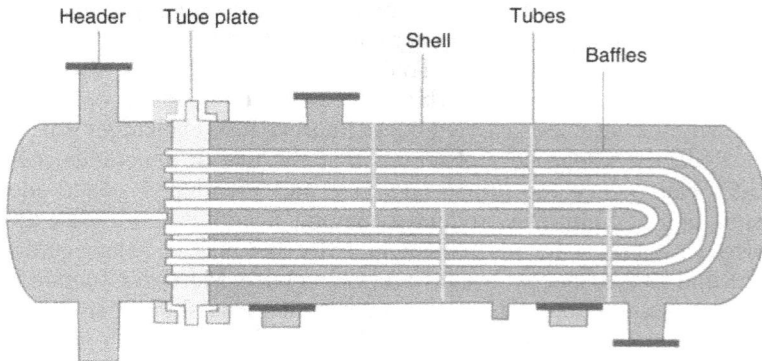

Figure 9.2 Cross-sectional view of shell and tube heat exchanger.
Source: Okonkwo et al. (2020)

fluid circulates between the tubes, while the fluid to be heated circulates within the shell through inlet and outlet valves. While these heat exchangers offer higher heat transfer rates due to their increase area of contact but they still face a limitation in terms of thermal conductivity when using conventional base fluids. Consequently, the substitution of these base fluids with nanofluids serves as an effective solution to overcome the challenge posed by low thermal conductivity.

The performance of a shell and tube heat exchanger consisting of Fe_2O_3-based water and ethylene glycol nanofluids was evaluated to access improvement in heat transfer rates by Kumar and Sonawane (2016). A study reported increments in overall and convective heat transfer coefficients that was directly proportional to NP concentrations. Enhancement in pressure drop was reported to be more prevalent under turbulent conditions, while, for laminar flow, a non-significant change in the pressure drop was reported.

The effect of CuO, Al_2O_3, ZnO and Fe_3O_4 NPs at 0.03% concentration with distinct mass flow rates in water-based nanofluids was evaluated by Shahrul et al. (2014) to fins the efficacy of shell and tube heat exchanger. The authors suggested that at a steady mass flow of 0.83 kg/s, the lowest and highest efficacy corresponded to Al_2O_3 and ZnO nanofluids, at 31.29 and 43.73%, respectively. Concerning the heat transfer coefficient, Al_2O_3 nanofluid exhibited a higher value at 12.06%. Thus, to improve the energy productivity of shell and tube heat exchangers, low tube side mass flow and high shell side mass flow rates should be used with nanofluids.

Modest heat exchangers containing two concentric tubes are generally known as double-pipe heat exchangers. Fluid flow in the geometry of double pipe heat exchangers could be concurrent (fluids flow in the same direction) or counter-current (fluids flow in different directions), as portrayed in Figure 9.3. Several studies have been undertaken to improve

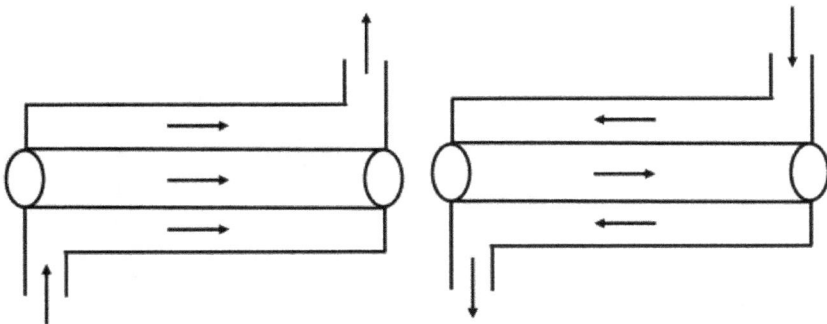

Figure 9.3 Double pipe type heat exchanger: (A) counter flow and (B) parallel flow.

Source: Gomez (2017)

the efficacy of heat exchangers, for example, modifications in dimension such as using corrugated or coiled tubes, bigger systems and extra-potent pump usage. Enhancement in the thermal performance of double tube heat exchangers with the application of various nanofluids has also been investigated by numerous researchers.

The impact of TiO_2 nanoparticles in ethylene glycol and water as a base fluid on friction factor and heat transfer coefficient was studied by Reddy and Rao (2014) in a double pipe heat exchanger with and without helical coil inserts. A study reported the Nusselt number was increased by 7.85 and 10.73% using a 0.02% volume fraction of TiO_2 nanofluid in comparison with the base fluid at Reynolds numbers of 4000 and 15,000, respectively, while with the insertion of helical coils, Nusselt number was found to increase by 16.11 and 17.71% compared to base fluid at Reynolds numbers of 4000 and 15,000, respectively. Furthermore, the study revealed that the presence of helical coils also led to an increment in the friction factor, with an increase of 13.39% and 16.58% at Reynolds numbers of 4000 and 15,000, respectively.

The effect of Al_2O_3 nanofluid in a mini-double pipe heat exchanger under turbulent conditions on the heat transfer performance was also determined by Aghayari et al. (2015). The authors reported that at a concentration of 0.2% and flow rate of 0.06 Kg/s, augmentation in heat transfer rates were observed around 7.32%, while at the concentration of 0.1%, a 1.71% increase in thermal efficiency was observed. For laminar flow conditions, Hussein (2017) observed elevation in the Nusselt number and friction factor at 35 and 12.5%, respectively, using aluminium nitride-ethylene glycol as a base fluid in a double tube heat exchanger. The outcome demonstrates that an increase in the heat transfer coefficient and friction factor is directly proportional to the concentration of NPs in the base fluid.

9.6.2 Plate Heat Exchanger

Plate heat exchangers (PHEs) are compact heat exchangers that have been functionalized in food processing units. In food processing, PHEs are most commonly seen in pasteurization processes. A plate heat exchanger allows for flow, in alternate channels, of the fluid to be processed and the thermal fluid, separated by a series of parallel plates with portholes through which the heat transfer between two fluids takes place.

Figure 9.4 illustrates a cross-sectional view of a plate heat exchanger. The geometry of the plate heat exchanger consists of one movable cover and one fixed cover, which is further composed of several heat transfer plates placed between a fixed plate and a loose pressure plate, forming a whole unit. Each heat transfer plate has a gasket arrangement that seals the channel and enables the primary and secondary media to flow in a counter-current

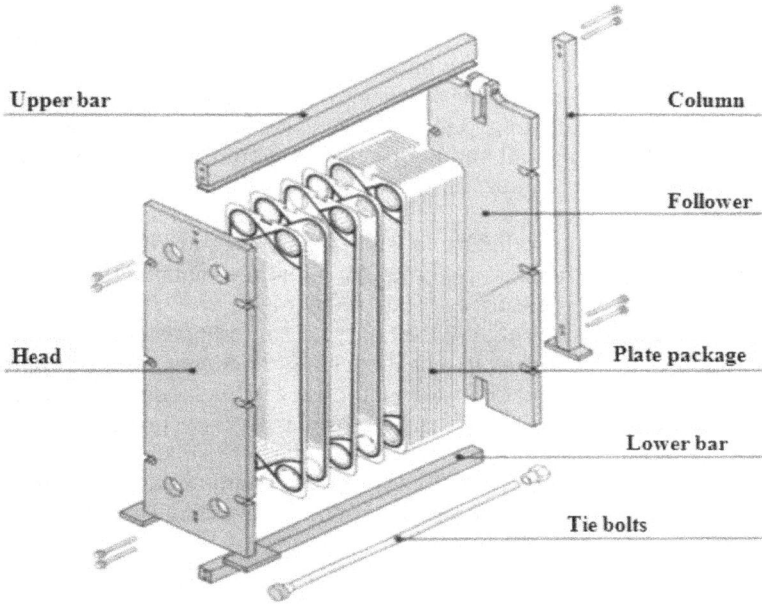

Figure 9.4 Cross-sectional view of plate heat exchanger.

Source: Fábio et al. (2015)

manner without being mixed, as shown in Figure 9.3. Plates of the heat exchanger are often corrugated to increase turbulence and achieve a higher heat transfer coefficient. Upper and lower carrying bars are also attached throughout the plates that support the channels and pressure between the plates. However, there is always room for improvement in thermal performance and efficiency; therefore, the employment of nanofluids offers high heat transfer rates within the same heat exchanger.

Several studies on the efficiency of nanofluids in PHEs have been conducted using different NPs (Sajid and Ali, 2019). Sajid and Ali observed an improvement in thermal heat performance of PHEs in nearly all studies they reviewed. Deteriorating effects, seen in limited studies, were attributable to NP agglomeration. The impact of viscosity at higher concentrations was also noted, adversely affecting thermal performance due to increased rates of sedimentation. Orientation and plate spacing also impacted PHE performance.

Mare et al. (2011) evaluated the use of alumina (Al_2O_3) and carbon nanotube (CNT) nanofluids, while Barzegarian et al. (2016) explored TiO_2 nanofluids. Both studies showed significant improvement in heat transfer characteristics. While Barzegarian's (2016) study showed no significant change in pressure drop, Mare et al.'s (2011) study showed superior pressure drop performance for both nanofluids (7× for CNTs and 3× for Al_2O_3) in comparison to water. The performance of TiO_2 water-based nanofluids in the mass fraction range of 0.3 to 0.15% was also investigated. The authors reported increased overall and convective heat transfer rates by 23.7 and 6.6%, respectively, in comparison to pure water (Barzegarian et al., 2016). Javadi et al. (2013) compared the heat transfer coefficients and pressure drops in a variety of nanofluids and found that Al_2O_3 and TiO_2 nanofluid had similarly high heat transfer properties, though these nanofluids also exhibited more drop with pressure as compared to SiO_2.

Taghizadeh-Tabari et al. (2016) carried out milk pasteurization using TiO_2/water nanofluids in a plate heat exchanger and observed an improvement in the convective heat transfer coefficient with increase in NP levels and Pe number. Conversely, it led to an increase in viscosity, pressure drop and pumping power. To evaluate the overall thermal performance, the researchers created a measure to calculate thermal economic efficiency called the thermal performance index, a useful expression given by Equation (7.1).

$$\eta = \left(h_{nf} / h_{nf} \right) / \left(\Delta P_{nf} / \Delta P_{bf} \right) = RE_h / RE_{\Delta P} \tag{9.8}$$

Calculated values were found to be greater than 1 under all flow conditions (in terms of Reynolds number). The highest value was 1.18 for TiO_2 at 0.8% concentration, indicating that the use of the nanofluid would be economical compared to water in terms of thermal efficiency.

Okonkwo et al. (2021) reviewed the use of hybrid nanofluids in PHEs. The maximum heat transfer augmentation of 27% was observed for Al_2O_3-TiO_2/water nanofluid. Improvement in the heat transfer rates and pressure drop was also investigated using different hybrid nanofluids within the heat exchanger. Bhattad et al. (2018) reported a 39.6% increase in the heat transfer coefficient in PHE using hybrid MWCNT-Al_2O_3 water-based nanofluids as compared to Al_2O_3 water-based nanofluid. Besides elevation in the heat transfer rates, pumping power could also be improved significantly by the use of hybrid nanofluids.

9.7 Impact on Various Food Properties

9.7.1 Physical and Chemical Properties of Foods

Application of nanofluids in heat exchangers for processing of liquid foods displays significant changes as compared to conventional fluids used for processing. Visual colour is one of the important characteristics that immediately affects the consumer. Since L^*, a^* and b^* values in the CIE system are used to gauge lightness, red-greenness and yellow/blueness, it is vital to assess the impact of innovative processing techniques on these colour parameters or values. Several authors measured the usage of nanofluids in heat exchangers and its effect on colour parameters from a consumer acceptance point of view.

Watermelon juice showed a greater retention of redness (a^*) in juice processed in a shell and tube heat exchanger with alumina-based nanofluids at 85°C for 45 sec, while a greater drop in redness was observed for juice processed in a heat exchanger with water at the same temperature and time. The reduction of duration of thermal processing due to nanofluids was found to be effective in the retention of redness of watermelon juice (Jafari et al., 2017). Furthermore, an increase in the b^* value of watermelon juice from 0.12 to 2.06 was observed with water at 75°C for 45 sec, while there was a significantly lower increase, 1.54, at the same temperature and time for juice processed with 4% alumina nanofluids (Jafari et al., 2017).

A few scholars studied the a^*/b^* (red green/yellow blue) colour index to express colour changes in commercial products. Further researchers observed that a^*/b^* increased with a higher percentage of red colour; hence it has been used to study redness of red fruits and vegetables as well as for meat (Pathare et al., 2013). The a^*/b^* index was significantly lower for tomato juice processed through a heat exchanger using water and comparatively low for tomato juice processed through a heat exchanger with 4% alumina nanofluids. Similarly, for watermelon juice, raising the concentration of NPs from 0 to 4% and reducing the process time from 45 to 15 sec elevated the value of a^*/b^* for all selected variables (temperature, duration and concentration).

The total colour difference is the difference between initial colour values and the actual colour coordinates and is generally denoted by ΔE^* (Mokrzycki and Tatol, 2011). Watermelon juice processed under normal thermal processing has a greater ΔE^* value than juice samples processed or pasteurized using alumina nanofluids. Greater values for ΔE^* are associated with dark colour of a sample. Basically carotenoids and lycopene pigments are responsible for the redness of watermelon, while their degradation leads to changes in a^* and b^* and further dark or brownish colouration in the watermelon juice (Klong-Klaew et al., 2018). Moreover, greater ΔE^* values for thermally processed juice samples are also associated with brown pigment formation during heating (Burdurlu et al., 2006).

The pH range of watermelon and tomato juice after processing was narrow, which indicated only a minor difference compared to fresh juice (Jafari et al., 2017). Furthermore, no significant difference in pH value at varying concentrations of NPs was observed. Similarly, the total soluble solids (TSS) index was also in range for processed and unprocessed watermelon and tomato juices. Therefore, it can be stated that thermal processing through nanofluids maintained the natural pH and TSS of juices after processing (Jabari et al., 2017; Jafari et al., (2017).

9.7.2 Nutrients and Various Functional Components

Vitamin C is one of the essential nutrients found in fresh fruits and vegetables and is essential for immune system functioning. It is also known as L-ascorbic acid, and it is a water-soluble antioxidant found in cells with strong anti-reactive oxygen species defences. L-ascorbic acid is oxidized to the unstable dehydroascorbic acid (DHAA) during heat processing, which is followed by hydrolysis, lactone ring opening and the production of the inactive compound 2,3-diketo-gluconic acid (Tola and Ramaswamy, 2015). Greater vitamin C retention was observed for watermelon and tomato juice (Table 9.2) processed through heat exchangers containing nanofluids as a heating medium. Moreover, it can be also noted that time and concentration of NPs have less effect on vitamin C retention than the temperature of processing; therefore, vitamin C retention can be improved at lower temperatures and times and higher NP concentrations (Jafari et al., 2017).

Phenolic compounds are secondary metabolites that received considerable attention because of their redox properties and physiological functions, namely antioxidant, anti-mutagenic and antitumor activities (Vinhas et al., 2021). A reduction in total phenolic content (TPC) was observed for tomato juice samples treated in a heat exchanger with water as a heating medium, while there was greater retention of TPC in samples processed through heat exchangers containing nanofluids as a heating medium. Additionally, for samples processed through heat exchangers using nanofluids as a heating medium, TPC was maintained at higher rates at longer processing times without drastically decreasing. This improved retention of TPC may be attributable to the shorter time required to thermally process the juice in the heat exchanger (Jafari et al., 2017).

Lycopene ($C_{40}H_{56}$), which gives fruits and vegetables like watermelon, tomatoes and others their red colour, has 2 unconjugated double bonds and 11 conjugated double bonds. According to epidemiological research, lycopene has diuretic and anti-inflammatory qualities and lowers the risk of heart disease and cancer. Therefore, from a nutritional perspective, one of the key goals is to maintain lycopene in processed juices. A drastic reduction in lycopene content was observed for watermelon and tomato juice processed through a heat exchanger containing water as a heating medium. Moreover, greater temperature of processing also affected the lycopene content of tomato juice (Table 9.2). A constant concentration of NPs (4%)

Table 9.2 Nanofluids as Heating Media and Their Impact on Various Food Properties

Food Sample	Nanofluid Composition	NP Diameter (nm)	NP Concentration (%)	Colour a* or a*/b**	Colour ΔE*	Total Soluble Solids (TSS)	Total phenolic Compounds Rate/ Lycopene Retention (%)*	pH	Vitamin C Retention (mg/100 mL)	Thermal Processing Time Reduction (%)	Thermal Processing System and Operating Conditions	Reference
Water-melon juice	Al₂O₃–water	50	unprocessed	2.18	–	9.1 to 9.2	–	5.48 to 5.60	–	–	Shell and tube heat exchanger/ 75, 80 and 85°C/15, 30 or 45 s	Jafari et al. (2017).
			0	1.27	3.26		81.15*		61.11	–		
			2	1.33	2.21		84.81*		63.70	24.88		
			4	1.46	1.14		91.28*		67.04	51.63		
Tomato juice	Al₂O₃–water		0	–	–	5.4 to 5.6	71.9		62.73	–	Shell and tube heat exchanger/ 70 for 30s	Jafari et al. (2017).
			2	–	–		–		65.59	22.23		
			4	–	–		73.6		66.34	46.29		
Tomato juice	Al₂O₃–water	20	unprocessed	2.2**	–	–	–	4.4 to 4.6-		–	Heat exchanger at 80°C	Jabbari et al. (2018).
			0	1.7**			66.81*			–		
			2	1.8**			–			22.2		
			4	1.9**			96.3*			46.3		

in nanofluids for heat exchangers at 75°C for 15 s for water melon juice and 70°C for 30 s for tomato juice showed 91% and 96% retention of lycopene. The greater retention of lycopene could be attributed to the more rapid heat transfer and lower processing time for the watermelon and tomato juice (Jafari et al., 2017; Jabbari et al., 2018).

Lengthy processing times in traditional pasteurization processes for fruit and vegetable juices generally cause nutrient loss and discoloration of juices. Moreover, reduction in processing time has importance in the fruit or vegetable juice industry, as it can reduce the time and energy consumption. Researchers observed a significant drop in processing time due to the usage of nanofluids as a heating medium compared to traditional fluids (generally water) for pasteurization. It was observed that watermelon and tomato juice processing using nanofluids as a heating medium reduced the processing time drastically. Additionally, greater concentrations of NPs (4%) used as a heating medium reduced total processing time to half for watermelon and tomato juice processing, as listed in Table 9.2.

Heat transfer properties of conventional fluids used for heating are responsible for process time variation and heat transfer rates. Furthermore, data in Table 9.2 reveal that the properties of heat transport are influenced by the concentration of NPs. The ability of nanofluids to transport heat depends on their capacity to do so, as well as their thickness, volume ratio, microstructure (morphology and size), thermal conductivity and other factors. It could be caused by a rise in the heat transfer coefficient of nanofluids, which improves thermal conductivity and encourages rapid dispersion of suspended particles and their interfacial movement, limiting layer growth at the edge (Keblinski et al., 2002).

9.8 Advantages and Limitations

Combining metal-based NPs in the base fluids of food processing equipment can enhance the heat transfer coefficient and hence the heat transfer rate. Nanofluids therefore heighten the efficiency of the food processing equipment mainly used in thermal processing and can preserve nutritional value through retaining essential nutrients and bioactive components at their highest concentration while minimally affecting the physical components such as colour, texture and sensory parameters (Jafari et al., 2017; Jabbari et al., 2018). Moreover, they are cost effective since they can shorten the heating time and energy consumption for processing.

The limitations regarding nanofluid industrial application in food sector is their stability in use. Basically, the sedimentation and aggregation of NPs affect their usage industrially. Another limitation that restricts their usage in thermal processing unit is that they need more pumping power for functioning since NPs increase the viscosity of nanofluids. Moreover, erosion and clogging of passages was also observed when nanofluids were used as a heat transfer fluid (Assael et al., 2019).

9.9 Future Perspectives and Conclusion

In the previous part of the chapter, we figured out the theory, principles, method of preparation, advantages and disadvantages of nanofluids. A considerable heat transfer enhancement was shown in thermal applications, that is, heating and cooling with the use of nanofluids. In principle, nanofluids possess improved thermal conductivity over base liquids. Stability is a major concern when using nanofluids in industrial applications; hence several researchers have suggested using surfactants to stabilize nanofluids, as stabilizers affect nanofluid droplet physical properties and henceforth their dynamic and heat transfer behaviour.

In the later part of the chapter, the different types of equipment used in the food industry that are chosen for nanofluids and their operations are discussed. Further, the effect of nanofluids on physicochemical and nutritional properties of food samples and their retention was discussed. The rheological behaviour of nanofluids and their prospects in heat transfer enhancement should also be investigated further for application in novel processing technologies such as high pressure processing, retort and cryogenic freezing.

Recently researchers have also studied hybrid nanofluids, where dissimilar NPs are suspended in a mixture or compound formulation to obtain the synergistic effect of suspended NPs. Broad experimental research should still be conducted to investigate their food application in the near future. Indeed, NPs diverge in morphology and geometry; therefore, for future investigation, these parameters should be considered in relation to their impact on the thermo-physical and functional characteristics of nanofluids. Moreover, studies on the influence of nanofluids as coolants in the food industry and their utilization as a heating medium for frying food samples in fast-food chains should be done to overcome the existing challenges.

List of Symbols:

A = Surface area
d = Thickness of layer
D = Channel diameter
d_p = Nanoparticle diameter
F = Tangential shear force
K = Conductivity constant
K_{bf} = Thermal conductivity of base fluid

K_{nf} = Thermal conductivity of nanofluid
K_p = Thermal conductivity of particle
n = Empirical constant, which is a shape factor
Nu_b = Nusselt number of base fluid
Pr_b = Prandtl number
Q = Amount of heat transferred
Re_b = Reynolds number of base fluid
RE_h = Ratio of heat transfer enhancement as compared with base fluid
$RE_{\Delta P}$ = Ratio of pressure drop of nanofluid as compared with base fluid
ΔT = Temperature gradient
μ = Dynamic viscosity
η = Intrinsic viscosity for a monodisperse system
ϕ = Nanoparticle volume fraction
ϕ_m = Maximum particle volume fraction at which flow can occur

References

Aghayari, R., Maddah, H., BaghbaniArani, J., Mohammadiun, H., & Nikpanje, E. (2015). An experimental investigation of heat transfers of Fe_2O_3/water nanofluid in a double pipe heat exchanger. International Journal of Nano Dimension, 6(5), 517–524.

Alia, B. S., Fryer, P., & Lopez Quiroga, E. (2019). Mapping energy consumption in food manufacturing. Trends in Food Science & Technology, 86. doi:10.1016/j.tifs.2019.02.034.

Assael, M. J., Antoniadis, K. D., Wakeham, W. A., & Zhang, X. (2019). Potential applications of nanofluids for heat transfer. International Journal of Heat and Mass Transfer, 138, 597–607.

Banisharif, A., Aghajani, M., van Vaerenbergh, S., Estellé, P., & Rashidi, A. (2020). Thermophysical properties of water ethylene glycol (WEG) mixture-based Fe3O4 nanofluids at low concentration and temperature. Journal of Molecular Liquids, Elsevier, 302, 112606. doi:10.1016/j. molliq.2020.112606. ⟨hal-02493762⟩.

Barzegarian, R., Moraveji, M. K., & Aloueyan, A. (2016). Experimental investigation on heat transfer characteristics and pressure drop of BPHE (brazed plate heat exchanger) using TiO2–water nanofluid. Experimental Thermal and Fluid Science, 74, 11–18.

Bhattacharjee, S. (2016). DLS and zeta potential—What they are and what they are not? Journal of Controlled Release, August 10, 235, 337–351. doi:10.1016/j.jconrel.2016.06.017. Epub 2016 Jun 10.

Bhattad, A., Sarkar, J., & Ghosh, P. (2018). Discrete phase numerical model and experimental study of hybrid nanofluid heat transfer and pressure drop in plate heat exchanger. International Communications in Heat and Mass Transfer, 91, 262–273.

Bhattad, A., Sarkar, J., & Ghosh, P. (2019). Energetic and exegetic performances of plate heat exchanger using brine-based hybrid nanofluid for milk chilling application. Heat Transfer Engineering, 41, 522–535. doi.org/10.1080/01457632.2018.1546770.

Biriukov, D., Fibich, P., & Predota, M. (2020). Zeta potential determination from molecular simulations. The Journal of Physical Chemistry C. doi:10.1021/acs.jpcc.9b11371.

Buongiorno, J. (2006). Convective transport in nanofluids. Journal of Heat Transfer, 128, 240–250. doi:10.1115/1.2150834.

Burdurlu, H. S., Koca, N., & Karadeniz, F. (2006). Degradation of vitamin C in citrus juice concentrates during storage. Journal of Food Engineering, 74(2), 211–216.

Carrillo-Berdugo, I., Midgley, S., Grau-Crespo, R., Zorrilla, D., & Navas, J. (2020). Understanding the specific heat enhancement in metal-containing nanofluids for thermal energy storage: Experimental and Ab initio evidence for a strong interfacial layering effect. ACS Applied Energy Materials, 3, 9246–9256. doi:10.1021/acsaem.0c01556.

Carvalho, P. M., Felício, M. R., Santos, N. C., Gonçalves, S., & Domingues, M. M. (2018). Application of light scattering techniques to nanoparticle characterization and development. Frontiers in Chemistry, June 25, 6, 237. doi:10.3389/fchem.2018.00237. PMID: 29988578; PMCID: PMC6026678.

Chen, J., Wang, F., Huang, K., Liu, Y-N., & Liu, S. (2009). Preparation of Fe 3O 4 nanoparticles with adjustable morphology. Journal of Alloys and Compounds, 475, 898–902. doi:10.1016/j.jallcom.2008.08.064.

Choi, S. U. S., & Eastman, J. (1995). Enhancing thermal conductivity of fluids with nanoparticles. Proceedings of the ASME International Mechanical Engineering Congress and Exposition, 66.

Choi, S. U. S., Zhang, Z. G., Yu, W., Lockwood, F. E., & Grulke, E. A. (2001). Anomalous thermal conductivity enhancement in nano-tube suspensions. Applied Physics Letters, 79, 2252–2254.

Curd, E. F., Railio, J., Gustavsson, J., Hogeling, J., El Haj Assad, M., Emilsen, J., Mazzacane, S., & Wiksten, R. (2001). 9—Air-handling processes. Editor(s): Howard, G., & Esko, T., Industrial Ventilation Design Guidebook, pp. 677–806, Academic Press. https://doi.org/10.1016/B978-012289676-7/50012-6.

Dhand, C., Dwivedi, N., Loh, X. J., Ng, A., Verma, N., Beuerman, R., Lakshminarayanan, R., & Ramakrishna, S. (2015). Methods and strategies for the synthesis of diverse nanoparticles and their applications: A comprehensive overview. RSC Advances, 5. doi:10.1039/C5RA19388E.

Esfe, M. H., Saedodin, S., Mahian, O., & Wongwises, S. (2014). Heat transfer characteristics and pressure drop of COOH-functionalized DWCNTs/water nanofluid in turbulent flow at low concentrations. International Journal of Heat and Mass Transfer, 73, 186–194. doi:10.1016/j.ijheatmasstransfer.2014.01.069.

237

Fábio, A. S., Mota, E. P. C., & Ravagnani, M. A. S. S. (2015). Modeling and design of plate heat exchanger, heat transfer studies and applications. Salim NewazKazi, IntechOpen. doi:10.5772/60885. www.intechopen.com/chapters/48647.

Ghadimi, A., Saidur, R., & Metselaar, H. S. C. (2011). A review of nanofluid stability properties and characterization in stationary conditions. International Journal of Heat and Mass Transfer, 54(17–18), 4051–4068.

Ghozatloo, A., Rashidi, A., & Shariaty-Niassar, M. (2014). Convective heat transfer enhancement of graphene nanofluids in shell and tube heat exchanger. Experimental Thermal and Fluid Science, 53, 136–141.

Gomez, A. (2017). Thermal performance of a double-pipe heat exchanger with a koch snowflake fractal design. Master of Science in Applied Engineering Thesis, Georgia Southern University.

Hamilton, R. L., & Crosser, O. K. (1962). Thermal conductivity of heterogeneous two-component systems. Industrial & Engineering Chemistry Fundamentals, 1(3), 187–191.

Hummers, W. S. & Offeman, R. E. (1958). Preparation of graphitic oxide. Journal of the American Chemical Society, March 20, 80(6), 1339. doi:10.1021/ja01539a017.

Hussein, A. M. (2017). Thermal performance and thermal properties of hybrid nanofluid laminar flow in a double pipe heat exchanger. Experimental Thermal and Fluid Science, 88, 37–45.

Jabbari, S-S., Jafari, S. M., Dehnad, D., & Shahidi, S. A. (2018). Changes in lycopene content and quality of tomato juice during thermal processing by a nanofluid heating medium. Journal of Food Engineering, 230, 1–7. doi:10.1016/j.jfoodeng.2018.02.020.

Jafari, S. M., Jabari, S. S., Dehnad, D., & Shahidi, S. A. (2017). Heat transfer enhancement in thermal processing of tomato juice by application of nanofluids. Food and Bioprocess Technology, 10(2), 307–316.

Jafari, S. M., Saremnejad, F., & Dehnad, D. (2017). Nano-fluid thermal processing of watermelon juice in a shell and tube heat exchanger and evaluating its qualitative properties. Innovative Food Science & Emerging Technologies, 42, 173–179.

Jama, M., Singh, T., Gamaleldin, S. M., Koc, M., Samara, A., Isaifan, R. J., Atieh, M. J. Critical review on nanofluids: Preparation, characterization, and applications. Journal of Nanomaterials, 2016. Article ID 6717 624, 22 pages, 2016. https://doi.org/10.1155/2016/6717624.

Javadi, F., Sadeghipour, S., Rahman, S., BoroumandJazi, G., Rahmati, B., Elias, M. M., & Sohel, M. R. (2013). The effects of nanofluid on thermophysical properties and heat transfer characteristics of a plate heat exchanger. International Communications in Heat and Mass Transfer, 44, 58–63. doi:10.1016/j.icheatmasstransfer.2013.03.017.

Kaggwa, A., Carson, J. K. (2019). Developments and future insights of using nanofluids for heat transfer enhancements in thermal systems:

A review of recent literature. International Nano Letters, 9, 277–288. https://doi.org/10.1007/s40089-019-00281-x.

Keblinski, P., Phillpot, S. R., Choi, S. U. S., & Eastman, J. A. (2002). Mechanisms of heat flow in suspensions of nano-sized particles (nanofluids). International Journal of Heat and Mass Transfer, 45, 855–863.

Khan, I., Saeed, K., & Khan, I. (2017). Nanoparticles: Properties, applications and toxicities. Arabian Journal of Chemistry. doi:10.1016/j.arabjc.2017.05.011.

Klong-Klaew, N., Saeng-Arun, T., & Sai-Ut, S. (2018). Effect of high temperature short time pasteurization on availability of bioactive compounds and quality changes of watermelon juice during refrigeration storages. Innovation of Functional Foods in Asia (IFFA), 1.

Kumar, N., & Sonawane, S. S. (2016). Experimental study of Fe_2O_3/water and Fe_2O_3/ethylene glycol nanofluid heat transfer enhancement in a shell and tube heat exchanger. International Communications in Heat and Mass Transfer, 78, 277–284.

Kumar, P. M., Kumar, J., Tamilarasan, R., Sendhilnathan, S., & Suresh, S. (2015). Review on nanofluids theoretical thermal conductivity models. Engineering Journal, 19(1), 67–83.

Lu, G. W., & Gao, P. (2010). Emulsions and microemulsions for topical and transdermal drug delivery. Handbook of Non-Invasive Drug Delivery Systems, 59–94. doi:10.1016/b978-0-8155-2025-2.10003-4.

Maheshwary, P. B., Handa, C. C., Nemade, K. R. (2017). A comprehensive study of effect of concentration, particle size and particle shape on thermal conductivity of titania/water based nanofluid. Applied Thermal Engineering, 119. doi:10.1016/j.applthermaleng.2017.03.054.

Mare, T., Halelfadl, S., Ousmane, S., Estellé, P., Duret, S., & Bazantay, F. (2011). Thermal performances of nanofluids at low temperature in a plate heat exchanger. Experimental Thermal and Fluid Science, 35, 1535–1543. doi:10.1016/j.expthermflusci.2011.07.004.

Maxwell, J. C. (1873). A treatise on electricity and magnetism (Vol. 1). Clarendon Press.

Merckx, B., Dudoignon, P., Garnier, J. P., & Marchand, D. (2012). Simplified transient hot-wire method for effective thermal conductivity measurement in geo materials: Microstructure and saturation effect. Advances in Civil Engineering, 2012.

Mikolajczyk, A., Gajewicz, A., Rasulev, B., Schaeublin, N., Maurer-Gardner, E., Hussain, S., . . . Puzyn, T. (2015). Zeta potential for metal oxide nanoparticles: A predictive model developed by a nano-quantitative structure–property relationship approach. Chemistry of Materials, 27(7), 2400–2407. doi:10.1021/cm504406a.

Mokrzycki, W. S., & Tatol, M. (2011). Colour differenceΔ E-A survey. Machine Graphics & Vision, 20(4), 383–411.

Noroozi, M., Radiman, S., Zakaria, A., & Soltaninejad, S. (2014). Fabrication, characterization, and thermal property evaluation of silver nanofluids. Nanoscale Research Letters, 9(1), 1–10.

O'Hanley, H., Buongiorno, J., McKrell, T., & Hu, L. (2012). Measurement and model validation of nanofluid specific heat capacity with differential scanning calorimetry. Advances in Mechanical Engineering, January. doi:10.1155/2012/181079.

Okonkwo, E. C., Wole-Osho, I., Almanassra, I. W., Abdullatif, Y. M., & Al-Ansari, T. (2020). An updated review of nanofluids in various heat transfer devices. Journal of Thermal Analysis and Calorimetry, 1–56.

Okonkwo, E. C., Wole-Osho, I., Almanassra, I. W. et al. (2021). An updated review of nanofluids in various heat transfer devices. Journal of Thermal Analysis and Calorimetry, 145, 2817–2872. https://doi.org/10.1007/s10973-020-09760-2.

Parametthanuwat, T., Bhuwakietkumjohn, N., Rittidech, S., & Ding, Y. (2015). Experimental investigation on thermal properties of silver nanofluids. International Journal of Heat and Fluid Flow, 56, 80–90.

Park, J. S., Kihm, K. D., Kim, H., Lim, G., Cheon, S., & Lee, J. S. (2014). Wetting and evaporative aggregation of nanofluid droplets on CVD-synthesized hydrophobic graphene surfaces. Langmuir, 30(28), 8268–8275. https://doi: 10.1021/la404854z.

Pathare, P. B., Opara, U. L., & Al-Said, F. A. J. (2013). Colour measurement and analysis in fresh and processed foods: A review. Food and Bioprocess Technology, 6(1), 36–60.

Perrier, C., & Beroual, A. (2010). Experimental investigations on insulating liquids for power transformers: Mineral, ester, and silicone oils. Electrical Insulation Magazine, IEEE, 25, 6–13. doi:10.1109/MEI.2009.5313705.

Ray, D. R., Das, D. K., & Vajjha, R. S. (2014). Experimental and numerical investigations of nanofluids performance in a compact minichannel plate heat exchanger. International Journal of Heat and Mass Transfer, 71, 732–746.

Reddy, M. C. S., Varma, R. R., & Rao, V. V. (2014). Experimental investigation of relative performance of water based TiO 2 and ZnO nanofluids in a double pipe heat exchanger. i-Manager's Journal on Mechanical Engineering, 5(1), 6.

Sadeghinezhad, E., Mehrali, M., Rahman, S., Mehrali, M., Tahan, L. S., Akhiani, A., & Metselaar, H. (2016). A comprehensive review on graphene nanofluids: Recent research, development and applications. Energy Conversion and Management, 111, 466–487. doi:10.1016/j.enconman.2016.01.004.

Sajid, M. U., & Ali, H. M. (2019). Recent advances in application of nanofluids in heat transfer devices: A critical review. Renewable and Sustainable Energy Reviews, Elsevier, vol. 103(C), 556–592.

Sarkar, J., Ghosh, P., & Adil, A. (2015). A review on hybrid nanofluids: Recent research, development and applications. Renewable and Sustainable Energy Reviews, 43, 164–177.

Selvamani, V. (2019). Stability studies on nanomaterials used in drugs. Characterization and Biology of Nanomaterials for Drug Delivery, 425–444. doi:10.1016/b978-0-12-814031-4.00015-5.

Seong, H. J., Kim, G. N., Jeon, J. H., Jeong, H. M., Noh, J. P., Kim, Y. J., Kim, H. J., & Huh, S. C. (2018). Experimental study on characteristics of grinded graphene nanofluids with surfactants. Materials, 11(6), 950. https://doi.org/10.3390/ma11060950.

Shahrul, I. M., Mahbubul, I. M., Saidur, R., Khaleduzzaman, S. S., Sabri, M. F. M., & Rahman, M. M. (2014). Effectiveness study of a shell and tube heat exchanger operated with nanofluids at different mass flow rates. Numerical Heat Transfer, Part A: Applications, 65(7), 699–713.

Sharif, M. Z., Azmi, W. H., Redhwan, A. A. M., Zawawi, N. N. M., & Mamat, R. (2017). Improvement of nanofluid stability using 4-step UV–Vis spectral absorbency analysis. Journal of Engineering Mechanics, SI 4(2), 233–247.

Silaipillayarputhur, K., & Khurshid, H. (2019). The design of shell and tube heat exchangers–A review. International Journal of Mechanical and Production Engineering Research and Development, 9(1), 87–102.

Simpson, S., Austin, S., Chris, G., & Saeid, V. (2019). Nanofluid thermal conductivity and effective parameters. Applied Sciences, 9(1), 87. https://doi.org/10.3390/app9010087.

Singh, L. P., Bhattacharyya, S. K., Kumar, R., Mishra, G., Sharma, U., Singh, G., & Ahalawat, S. (2014). Sol-Gel processing of silica nanoparticles and their applications. Advances in Colloid and Interface Science, December, 214, 17–37. doi:10.1016/j.cis.2014.10.007. Epub 2014 Nov 6. PMID: 25466691.

Taghizadeh-Tabari, Z., Zeinali, H. S., Moradi, M., & Kahani, M. (2016). The study on application of TiO2/water nanofluid in plate heat exchanger of milk pasteurization industries. Renewable and Sustainable Energy Reviews, 58, 1318–1326. doi:10.1016/j.rser.2015.12.292.

Tamilselvan, K., Sivabalan, B., Prakash, R., Manojprasath, M., & Mahabubadsha, A. (2017). Experimental analysis of heat transfer rate in corrugated plate heat exchanger using nanofluid in milk pasteurization process. International Journal of Engineering and Applied Sciences, 4(5), 257465.

Tola, Y. B., & Ramaswamy, H. S. (2015). Temperature and high pressure stability of lycopene and vitamin C of watermelon Juice. African Journal of Food Science, 9(5), 351–358.

Tripathy, D. B., & Mishra, A. (2016). Sustainable biosurfactants. Sustainable Inorganic Chemistry, 1, 175–192.

Tyndall, J. (1868). On the blue colour of the sky, the polarization of sky-light, and on the polarization of light by cloudy matter generally.

Proceedings of the Royal Society of London, 17, 223–233. doi:10.1098/rspl.1868.0033.

Utomo, A. T., Poth, H., Robbins, P. T., & Pacek, A. W. (2012). Experimental and theoretical studies of thermal conductivity, viscosity and heat transfer coefficient of titania and alumina nanofluids. International Journal of Heat and Mass Transfer, 55(25–26), 7772–7781. doi:10.1016/j.ijheatmasstransfer.2012.08.003.

Vinhas, A. S., Silva, C. S., Matos, C., Moutinho, C., & Ferreira da Vinha, A. (2021). Valorization of watermelon fruit (*Citrullus lanatus*) byproducts: Phytochemical and biofunctional properties with emphasis on recent trends and advances. World Journal of Advance Healthcare Research, 5(1), 302–309.

Xian, H. W., Sidik, N. A. C., Aid, S. R., Ken, T. L., & Asako, Y. (2018). Review on preparation techniques, properties and performance of hybrid nanofluid in recent engineering applications. Journal of Advanced Research in Fluid Mechanics and Thermal Sciences, 45(1), 1–13.

Yu, W., & Xie, H. (2012). A review on nanofluids: Preparation, stability mechanisms, and applications. Journal of Nanomaterials, 2012, Article ID 4 35873, 17 pages, 2012. https://doi.org/10.1155/2012/435873.

Chapter 10

Nanobiosensors for Detection of Food Contaminants

Prastuty Singh and Antima Gupta

10.1 Introduction

Contaminants are substances that unintentionally reach food during its production, processing, marketing or distribution. With the rise in awareness of consumers towards health and safety, demand for food products free from various chemical, physical, radiation and biological contamination is increasing rapidly. Microbial contamination is a source of major concern for food products due to their high nutritional content. Microorganisms usually produce toxins as a means of self defense, contaminating food and ultimately serving as the basis for various diseases. Likewise, food adulteration is another cause of negative impacts on human health. Toxins and adulterants are capable of causing different health issues varying from diarrhea, vomiting and paralysis to even death. Food safety is considered a global concern due its primary requirement in human health. Therefore, it is essential to build sensitive and rapid detection techniques to identify unsafe contaminants such as chemicals, pesticides, bacteria and their toxins (Zhu et al., 2009). Typically, microbiological analysis includes different procedures like preparation of the sample, culturing and isolation in suitable media and detection and identification of obtained isolates. These microbiological

DOI: 10.1201/9781003217138-10

culturing techniques have already been standardized and offer appreciable specificity and sensitivity. Conversely, these traditional techniques are also very labor intensive and tedious and require relatively large quantities of materials for analysis. Elevated interest by consumers in food safety has led to the development of modern food processing techniques that are high throughput, cost efficient, reliable and rapid (Velusamy et al., 2010). At present, mounting interest in nanotechnology can easily be observed, as it has given rise to novel prospects and unique applications across various fields as well as the food industry. Consequently, nanotechnology has instigated fresh challenges for improvement and innovation in the food industry at a massive rate in the field of biosensing food properties.

Nanomaterials are classified as substances which at the minimum have one measurement smaller than 100 nm. Nanomaterials are divided on the basis of their dimensions: one dimension, coatings and nanofilms; two dimensions, wires and nanotubes; and three dimensions, nanoparticles (Hochella, 2002). Unique characteristics (chemical and physical) of nanomaterials are associated with their extremely small size and high surface area. Taking into account the current scenario and benefits of nanomaterials, nanosensors have utility in competent monitoring of different food contaminants with regard to pH, ingredients, microbiological count and humidity and therefore serve as a potential tool to upgrade the quality and safety of food materials. Nanosensors can be defined as sensory points (chemical, biological and surgical) being applied to transmit information regarding nanoparticles to a relatively larger scale (macroscopic) (Foster, 2006). Hence, nanosensors can be described as an altered type of biological or chemical sensor utilizing nanomaterials in the analysis apparatus.

The key difference between the two is that chemical sensors make use of chemical reagents, while biosensors utilize bio-reagents. On the basis of this, nanosensors are categorized as chemical nanosensors and nanobiosensors (Agrawal and Prajapati, 2012). Another criterion of classifying nanosensors is on the basis of their utility, where they are categorized as nanobiosensors, chemical nanosensors, deployable nanosensors and electrometers. However, the current chapter will focus on nanobiosensors and chemical nanosensors. The chapter discusses nanomaterial-based biosensors that have evolved in recent times for the diagnosis of various types of contamination and focuses mainly on their nanostructure, influence on food properties and crucial role in enhancing the performance of detection.

10.2 Novelty

Behavior analysis of non-constant materials is of immense importance in areas such as environmental application, screening food quality and pharmaceutical identification. In the given context, expansion of proficient biosensors that can assess the minor details of any interaction with great accuracy and utmost sensitivity requires top prioritization. A major constituent of biosensing techniques is transduction processes responsible for transforming bioanalyte interaction responses in a reproducible and detectable fashion by utilizing the transformation of energy from distinct biochemical reactions into an electrical form through a transduction process. Nanomaterials serve as a brilliant alternative in this task due to their high surface area to volume ratio, which permits the surface to be employed in a functional and efficient manner. Furthermore, the electromechanical attributes of nanomaterials are another useful quality for biosensors (Mustafa et al., 2020a).

10.3 Principles and Theory

A biosensor is a sensing system or measurement device created particularly for evaluation of a material or substance by employing biological interactions, followed by analyzing and converting the interactions into readable form by a transduction process and electromechanical interpretation. On the basis of mode of operation, a biosensor has three major components: bioreceptor, transducer and detector (Figure 10.1). Bioreceptor of a biosensor is used as a prototype for the material that needs to be identified. Various materials can be used as bioreceptors (Malik et al., 2013). For example, an antigen is separated using antibodies, and protein screening is done using its specific complementary substrate. A transducer system includes the second component, which aids in the conversion of interaction between a bioanalyte and its complementary bioreceptor to an electrical form (one form of energy to another). Due to the particular interaction between the bioreceptor and bioanalyte, the initial type of energy is biochemical, whereas the second form is typically electrical. The third constituent of a biosensor is a detector system. This component collects the electrical signals produced by the transducer and amplifies them appropriately so that the complementary

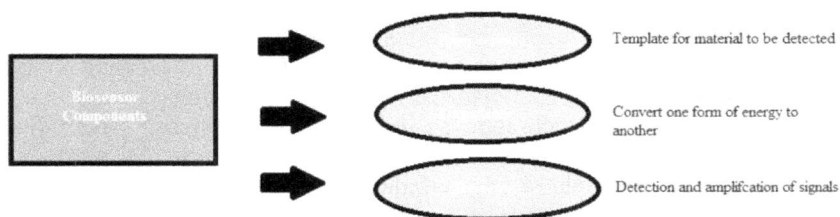

Figure 10.1 Three basic components of a biosensor.

response or reaction can be examined and interpreted properly. The concept that sets nanobiosensors apart from other biosensors is significant prerequisite of available immobilization schemes that are used for the purpose of bioreceptor immobilization in order to make bioanalyte reaction more practical and effective. The potential of nanobiosensing techniques is also influenced by variations in pH, temperature, contaminant interference and physicochemical changes (Kissinger, 2005).

10.4 Merging Nanotechnology with Biosensors

Nanomaterials have one of their dimensions under 100 nm, which usually makes them unique. Most of the associated elements of nanomaterials are located near or at their surface and possess all the essential physicochemical properties which differ prominently from the identical material at a bulk scale (Malik et al., 2013). Nanomaterial-based integrated devices with electrical systems produce nanoelectromechanical systems (NEMS), which are known for their exceedingly active electrical transduction processes. Nanocrystalline-based thin films, nanowires, nanotubes, nanoparticles and nanorods are a few of the materials being utilized due to their enhanced transduction process and biological signaling (Jianrong et al., 2004). Of these, the application of nanoparticles has been intensely analyzed and studied so far. The selection criteria of nanomaterials for use in biosensors are usually decided on the basis of output quantity of biological response along with transformation of generated biological response to perform further analysis and interpretation. For the application of biosensing techniques, the selection of nanomaterials is heavily reliant on their properties. Some widely used nanomaterials employed in sensor-based technologies are shown in Figure 10.2.

To enhance the biological response specificity of nanosensors, they are used at the same scale as biological process. Nanofabrication is one of the major operations of experimental design before integration of nanomaterials with the sensing apparatus. This step has two significant actions: utilization of an integrated circuit for the design and development of nanoscale adhesive and nanomaterial surface engineering by the use of micro-machining techniques. Photolithography, surface etching strategies, thin film etching/growth and chemical bonding are the four fundamental processes associated with nanofabrication-based biosensing (Gerwen et al., 1998). The lithography technique is used to produce nanoscale electrodes, which further provide more surface area for immobilization with greater sensitivity. With the help of enzyme glucose oxidase, this technique was first utilized to develop glucose biosensors. It was found that platinum nanoparticles added over carbon nanotube sheets enhanced detection of analytical material in biosensors, making identification of glucose possible

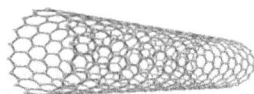

Carbon nanotubes (8–50 nm).

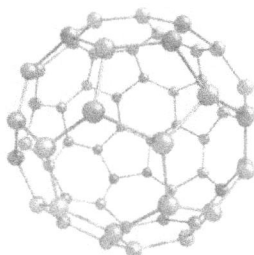

Gold nanoparticle (5–400 nm).

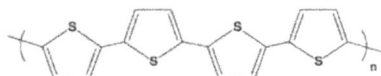

polythiophene

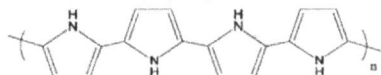

polypyrrole

polyaniline

Conducting polymers (20, 60, 100 nm).

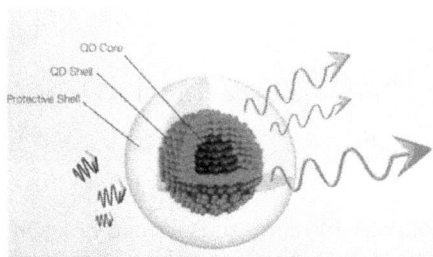

Magnetic beads (5–100 nm).

Figure 10.2 Various kinds of nanomaterials commonly used for the development of nanosensors.

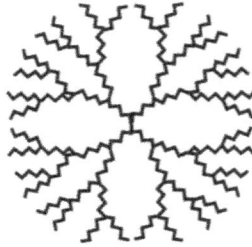

Dendrimers (2–20 nm).

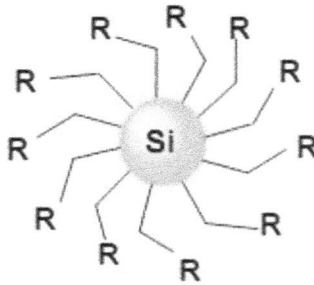

Silicon nanoparticles (3–10 nm).

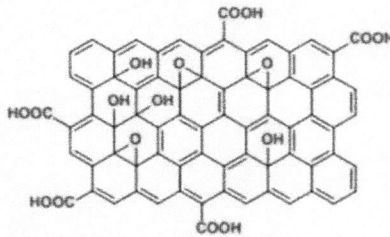

Graphene oxide (1.18–10.5 μm).

Figure 10.2 (Continued)

from a variety of sources. Immune sensors can be used to detect the anti-gen-antibody complex produced during the reaction where thin films are usually coated on sensing surface and allow better and quick identification of analytes (Pak et al., 2001). Nano-electromechanical systems possess excellent sensitivity with nanoscale functions and, when merged along with nanomaterials, provide unique new properties in sensing, portable

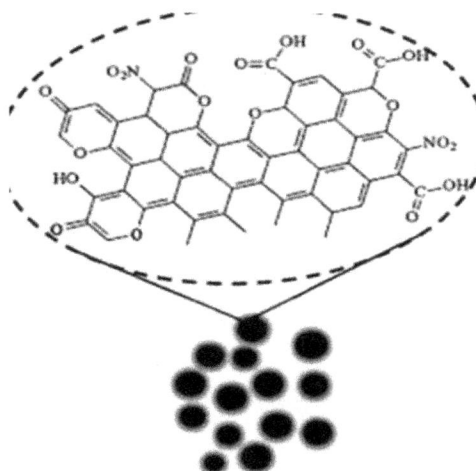

Carbon quantum dots (<10 nm).

Figure 10.2 (Continued)

generation of power, displaying and energy harvesting. When NEMSs are engineered with micro-electrochemical systems, it provides improved performance with mechanical materials and bio-adhesion to a variety of stimuli. Fluorophores, which are another significant component, absorb and emit light in the wavelength of emission and excitation spectra by making use of total internal reflection. These systems are employed as detection reagents along with antibodies in flow cytometry.

10.5 Types of Nanobiosensors

10.5.1 Carbon Nanotubes

In last few years, carbon-based nanotubes arose as one of the most utilized nanomaterials in biomolecular and drug delivery–based techniques. They are hollow cylindrical tubes consisting of one or more graphite-based layers surrounded by fullerenic hemispheres, known as single- to multiwalled carbon nanotubes (CNTs) (depending on graphite layers). They have high surface to volume ratio, distinct structure and electrocatalytic action, good chemical stability, high mechanical and electrical properties, light weight, high thermal conductivity and negligible surface fouling (Wang et al., 2017). Single-walled surface carbon nanotubes (SWCNTs) were

249

utilized for recognition of *Salmonella* in a DNA sensor–based application with the help of N-ethyl-N'-(3-dimethylaminopropyl) carbodiimide hydrochloride (3-dimethylaminopropyl) covalently bonded to the CNT (Weber et al., 2011). Sensors illustrated sensitivity at 1 mol/L DNA (target concentration) without any fluctuation. The use of CNT-immobilized glucose biosensors for glucose oxidase enzyme is very popular. Conventionally, glucose biosensors are only used to determine glucose from major tissues of the body, but recently, with the immobilization of nanotubes in the assembly, glucose can be estimated even from scarce body liquids like saliva and tears (Pak et al., 2001), making it quite essential for rapid detection using the lowest amount of sample.

10.5.2 Gold Nanoparticles

Gold nanoparticles (AuNPs) are popular in the nanotechnology field. They are typically synthesized in aqueous or organic solvents which require a surfactant (stabilizing mediator) for maintaining stability. Absorption of a suitable mediator or chemical bonding on gold NPs is exploited to attain surfactant-based stability of these nanoparticles. In order to avoid nanoparticle aggregation, the surfactant needs to be loaded first, and various properties of gold NPs can be standardized by the selection of different surfactants (Sperling et al., 2008). Unique conducting capacity, high surface to volume ratio and biological compatibility make AuNPs an extremely efficient nanoparticle (Guo and Wang, 2007). Redox activity is another fascinating feature of AuNPs that helps in improving the electrochemical biosensor sensitivity during foodborne pathogen analysis. Application of AuNPs onto electrochemical biosensor where AuNPs are in integration with ss DNA that is complementary to the microbial DNA under analysis leads to enhanced DNA-gold NPs based binding on the surface of transducer and further helps in improving biosensor sensitivity (Zhang, 2007). In a study conducted by Sperling et al. (2008), redox enzymes were added along with gold NPs that accurately reduce/oxidize the analyte as the substrate. On analyte binding, the AuNPs conjugated with enzymes enhanced the current signal and helped confirm detection. In a study conducted by Davis et al. (2013), an AuNP-modified electrode biosensor was employed to observe *L. monocytogenes* presence within spiked blueberry where the detection limit after 1 hour of analysis was found to be 2 log cfu/g. Another study documented the preparation of improved AuNP aptasensors by means of surface-enhanced Raman spectroscopy (SERS) in immediate identification *of S. typhimurium* and *S. aureus* in a pork spiked sample. On comparing the results of an AuNP aptasensor with other sensors, it was observed that detection limits for *E. coli, Bacillus cereus, V. parahaemolyticus* and *Shigella dysenteriae* were quite low (Zhang et al., 2015).

10.5.3 Quantum Dots

Carbon quantum dots (CDs) are minute, quasi-spherical (diameter < 10 nm) and artificial semiconductor particles. Due to their high biocompatibility, solubility and luminescence properties, they are being exploited in various research areas (Xu et al., 2004). Carbon quantum dots are nanoparticles developed from sp2 hybridization of graphite core (crystalline) and amorphous aggregates used in bio-labeling and bioanalytics. Conversely, graphene quantum dots (GQDs) are made up of a single or a few graphene lattices (usually less than 10). Because of their periodic structure and conjugated domain, GQDs typically possess a more crystalline nature than CDs. Such carbon-based nanostructures are two separate allotropes in reality, and both of them are functionalized using oxygen-linked complex surface group entities like derivatives from hydroxylate and carboxylate, improving the particle solubility and optical characteristics (Li et al., 2012). The difference in the formulation of such materials produces a variety of surface functionalization as well as complex hybridization with biosensor applications. For the detection of harmful toxins and microbial cultures such as *S. typhimurium* and aflatoxin B1 (AFB1), carbon quantum dots have been employed. Carbon dot aptamer complexes (CD-apts) were produced by Wang et al. (2015) for the quantitative detection of *S. typhimurium* in tap water solutions and eggshells at a limit of detection (LOD) of 50 cfu/ml, test range of 10^3 to 10^5cfu/ml. Different trials have confirmed the microbial identification–related specificity of CDs, carbon nanotubes, GQDs and semiconductor-based biosensors in both simple and complex settings of food. Wang et al. (2016) formulated AuNPs, CDs and aptamer-based detection systems for the specific identification of AFB1 and attained a LOD of 5 pg/ml. The same process was applied with varied supplemented concentrations of AFB1 in peanuts and corn samples, and results reported an average recovery of 92–105%. The obtained recovery improved by altering the aptamer with that of an ochratoxin aptamer which was fabricated to analyze only fungal toxins.

10.5.4 Magnetic Nanoparticle Beads

A new class of nanomaterials is composed of magnetic nanoparticles used in biosensors that can be modified by varying the magnetic field. These nanoparticles are made up of a magnetic bead cluster with a diameter in the range of 50–500 nm (Liu et al., 2011). Magnetic nanoparticles have gained popularity as a significant fabrication material for the purpose of developing a flow assay with a strong color that distinguishes the target material from a complex matrix. Such NPs have a supplementary advantage of offering strong visual and magnetic signals. Magnetic beads have been employed in various studies to fabricate flow assays that can facilitate detection of pathogenic bacteria. D.B. Wang et al. (2015) fabricated 300-nm-long magnetic beads covered with an antibody and utilized them for *Bacillus anthracis* spore identification with a detection limit of 5×10^5 spores/g of baking soda, 6×10^4 spores/g of milk powder and 2×10^5 spores/g of starch. The developed sensor proved its superiority, as it didn't illustrate any kind of specificity towards various other *Bacillus* species such as *B. thuringiensis, B. mycoides* and *B. cereus*. In comparison to traditional lateral flow process, this method does not necessitate sample pre-treatment and offers rapidity for naked eye, optical and magnetic identification within a time period of 20 minutes. Xia et al. (2016) made a rapid detection system for *S. choleraesuis* using gold magnetic nanobeads with a detection limit of 5×10^5cfu/ml in 20 hours for whole milk in contrast to a colloid gold-based lateral flow assay, which had higher detection limit of 5×10^6 cfu/ml, substantiating the supremacy of magnetic nanobeads to that of colloid gold.

10.5.5 Dendrimers

Dendrimers, or DEN, are globular-branched complex structures with a size within 2–20 nm. The preference given to DEN as synthetic nanoparticles for the development of biosensors is because of their various superior structural features such as convenient size, monodisperse nature, hydrophilicity, easily adaptable surface functionalities and high chemical as well as mechanical strength. The polyamidoamine also known as PAMAM dendrimer has gained major interest, as it grants the benefit of large surface areas with a good amount of functional groups in order to facilitate uncomplicated binding of biological bodies. At the end of the dendrimer structure, they also have monodisperse and hyper-branched polymers, including active functional groups. The immobilization of various bio-recognizing molecules is done with the help of these functional groups, which act as bio-conjugating components. Electrochemical technologies such as electro-chemiluminescent, potentiometric, impedimetric and amperometric technologies are typically used with dendrimers for determination of particular molecules with good sensitivity and selectivity (Bahadir and Sezgintürk, 2016). Shiddiky et al. (2007) fabricated a sandwich-based competitive bioassay for analyzing protein and DNA with the help of H_2O_2 reduction activity of poly-5,20:50,2"-terthiophene-30-carboxylic acid conducting polymer. Shiddiky et al. (2008) developed a biosensor using a 3G PAMAM dendrimer which was covalently linked to AuNP/CDS nanoparticles immobilized on bioreceptor molecule by chemisorption and illustrated a recognition limit of 450 aM and 4 fg/mL for DNA and protein, respectively. When compared with the plain pTTCA layer, the developed biosensor was found to be 70 times more sensitive due to the presence of AuNPs on the pTTCA/DEN layer that aided the linking of hydrazine, proteins and avidin.

10.5.6 Silicon Nanomaterials

Highly specific silicon nanostructure-based sensors are being fabricated with the purpose of rational fabrication bio-imaging and biosensing-based applications (Wang et al., 2013). Due to their biodegradable nature, silicon nanomaterials (SiNPs) can easily be transformed into renal clearance molecules that leave no toxic evidence on being excreted from the body (Park et al., 2009). The FDA has approved the application of SiNPs with diameters ranging from 3–10 nm in human clinical trials (Phillips et al., 2014). For the selection and precise identification of reproducible biological and chemical species, various silicon surface-enhanced Raman scattering or SERS-based sensors are being used. With the help of chemiluminescence assays, porous silicon-based biosensors are used for the detection of pathogenic strains of colony-forming units of *E. coli*. The sensitivity of such biosensors against *E. coli* was established to be 10^2 and 10^1 cells for 30 and 40 minutes, respectively. A study by Mathew and Alocilja (2005) employed the process of anodization that was used in an electrochemical Teflon cell for the development of porous silicon. A strain of *S. enteritidis* was taken as a probe, whereas a platinum wire and silicon chip were taken as cathode and anode, respectively. It was observed that porous silicon-based biosensors were extremely sensitive, with a greater surface in contrast to planar silicon-based biosensors and targeted DNA-specific probes.

10.5.7 Graphene-Based Nanomaterials

Another kind of material being utilized as a transducer in various biosensors is graphene-based nanomaterials. Usually such nanomaterials are employed for the transformation of receptor and targeted moieties for identifiable measurement with the help of EDC/NHS chemistry (Pumera, 2011). Conversely, graphene is considered the most widely exploited nanomaterial in the design of a variety of biosensors with various modes of transduction, as it possesses a high surface area and the capability to immobilize a variety of molecules with high electrical conductivity and rate of transmission (Rao et al., 2009). Graphene-based nanomaterials also find utility as quenchers to produce fluorescent transducer-based biosensors because graphene (G), graphene oxide (GO) and reduced graphene oxide (rGO) have significantly enhanced efficiency for fluorescent quenching. When designing sensors, graphene affects the detection limit of targeted

molecules and orientation of GO, rGO or G sheets along with bioreceptors, influencing the selectivity and sensitivity of biosensors (Kasry et al.,
2012). The oxidation state of graphene, functional groups, usage of various
derivatives and number of layers bring about changes in the sensing performance of nano-biosensors. For determination of *E. coli* O157:H7, with
the help of a specific probe oligonucleotide, Tiwari et al. (2015) fabricated
an electrochemical sensor which was based on GO-modified iron oxide-
chitosan hybrid nanocomposite on which the oligonucleotide was immobilized. The observed results illustrated that the sensor has detection limit of
1×10^{-14} to 1×10^{-16} M. On the other hand, pDNA/GIOCh/ITO bioelectrode
specificity with *E. coli, Klebsiella pneumonia, S. typhimurium* and *Neisseria
meningitides* samples and different target DNA sequences demonstrated
signals of insignificant value.

10.5.8 Conductive Polymers

Conductive polymers with unique properties provide an alternative to
material currently being used in fabrication of biosensors. Polymers are
generally considered good insulating materials in nature, but when some
polymers are combined with semiconductor and metallic characteristics, they show good conducting characteristics as well. Different conducting polymers are being used in a variety of applications (Faridbod
et al., 2008). Among these, polypyrrole, polythiophene and polyaniline
illustrate biocompatibility, decrease major disturbance sources impacting the working environment and prevent fouling of electrodes and thus
are employed as nanomaterials (Geise et al., 1991). For the purpose of
identifying foodborne pathogens, only polypyrrole and polyaniline are
extensively used. Due to high selectivity and sensitivity, superior transduction of signals, durability, flexibility and biocompatibility, conducting polymers are employed as a brilliant platform for immobilization of
biomolecules over electrodes (Malhotra et al., 2006). *L. monocytogenes*
cell-surface protein was identified by the use of a direct immunosensor
with label-free immunosensing of InIB (Internalin) with a detection limit
of 4.1 pg/ml (Tully et al., 2008). Another study reported electrochemical polyaniline transducer-based biosensor performance for estimating
an immune reaction for the detection of *E. coli* O157:H7 with a detection limit of 7.8 × 101 cfu/ml under 10 minutes (Muhammad-Tahir and
Alocilja, 2003). Muhammad-Tahir and Alocilja in (2004) reported a detection limit of 81 cfu/ml in 6 minutes for fresh produce like alfalfa sprouts,
strawberries and lettuce, and the use of polyaniline as an identifier has
been documented for detection of *E. coli* O157:H7-based electrochemical
sandwich immunoassay.

Table 10.1 Studies on Detection of Pathogens/Toxins Using Various Nanostructure-Based Sensors

Nanomaterial	Function	Pathogen/ Toxin	Importance	Reference
Gold nanoparticles	Rapid screening with high sensitivity	Bacterial decarboxylation of histidine resulted in histidine detection	Histidine presence is considered a spoilage marker for stored poultry meat	El-Nour et al. (2017)
Carbon nanotubes	On-state current of the functionalized nanotube sensor decreased sharply on exposure to *Salmonella*	Detection of *Salmonella* in nutrient broth	Food safety monitoring	Mitchell et al. (2011)
MNP-antibody-based impedimetric nanobiosensors	Increase in sensitivity to pathogen (by 35%)	Detection of *E. coli*	Detection of pathogens in food product	Khan et al. (2019)
Silicon Nanomaterials	Porous silicon-based nano-biosensors were highly sensitive with more active surface area	*S. enteritidis*	Detection of food-borne pathogens	Zhang and Alocilja (2008)
Graphene oxide/gold nanocomposite	More surface area with high conductivity led to more sensitivity and quick identification of patulin	Patulin	Detection of fungal-mycotoxin presence in food products, especially apple products	Song et al. (2021)

(Continued)

255

Table 10.1 (Continued)

Nanomaterial	Function	Pathogen/ Toxin	Importance	Reference
Hydroxy-terminated polyamido-amine dendrimer	PAMAM-OH presence in water results in permeation of SYTOX dye into the bacterial cell, enhancing the fluorescence significantly	*Pseudomonas aeruginosa*	Detection of *P. aeruginosa*, as it is cause of food poisoning and spoilage in food and beverages	Chang et al. (2001)
Polypyrrole-co-3-carboxyl-pyrrole copolymer (conductive polymer)	Excellent platform for immobilization with high selectivity for *S. typhimurium* over other pathogens	*S. typhimurium*	Detection of Gram-negative foodborne pathogens for consumer safety	Sheikhzadeh et al., 2016

10.6 Mode of Action of Nanostructures for Food Analysis

10.6.1 Nutrients (Antioxidants)

Antioxidants are widely known for their health beneficiary effects, as they prevent the accumulation of excessive free radicals, which has been associated with various diseases like aging, neurodegenerative issues, rheumatoid and cancer. In recent years nanosized materials with exceptional properties have been implemented for the quantification of antioxidants in food content with improved cost, sensitivity and portability (Valko et al., 2006). Most such analysis has been derived from silver- and gold-based nanostructures (nanorods, nanoparticles). Catalytic or conductive nanostructure-modified electrodes are being used to directly measure antioxidants at their distinguishable oxidoreduction potential. Other sensors biocatalytically transform antioxidants (specifically those with phenolic groups) by using enzymes such as tyrosinase and laccase to quinines, which are usually quantified at low applied voltage with electrochemical reduction

(~–100 mV) because of minimum interference from present electroactive species. On the other hand, nanomaterials like carbon nanotubes are incorporated with a matrix of enzyme immobilization to enhance sensitivity of detection towards polyphenolic compounds like trolox. This two-step biochemical and electrochemical process is illustrated by the subsequent reaction, where Q and HQ_2 are the oxidized and reduced forms of phenolic compounds (Mustafa and Andreescu, 2020b).

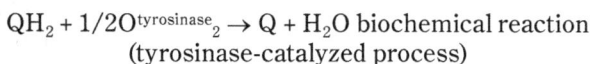

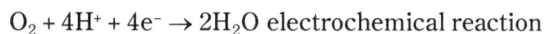

$$QH_2 + 1/2O^{tyrosinase}_2 \rightarrow Q + H_2O \text{ biochemical reaction}$$
(tyrosinase-catalyzed process)

$$O_2 + 4H^+ + 4e^- \rightarrow 2H_2O \text{ electrochemical reaction}$$

10.6.2 Contaminants

Heavy metals and pesticides present in food are a serious health hazard. Probable danger by leaching of various toxic substances from packaging materials is also dangerous. An example of such an incident is the chemical bisphenol A (BPA). BPA is utilized in the manufacture of epoxy resin and polycarbonate, which are further used to manufacture feeding bottles, water bottles and food cans. FDA banned the use of polycarbonate-based baby bottles for milk in 2012 (Andrei et al., 2016). So for detection of the electrochemically active chemical BPA, direct oxidation at the surface of electrodes can be done. To achieve this, carbon nanotubes or metal nanoparticle-modified electrodes can be employed to improve the detection limit. Conductive layer of multi-walled carbon nanotubes modified the glass carbon electrode along with the molecularly imprinted polymer (used as a particular BPA binding site). This further facilitated more sensitive and precise electrochemical determination of BPA and proved its utility in feeding bottles for infants for the detection of BPA.

10.6.3 Adulterants

To modify the sensory attributes of food products, a variety of exogenous compounds like dyes are used. Amongst adulterants, Sudan I, a food azo dye, is often used for the purpose of food coloring. More than 570 products with chili powder as an ingredient were found to be contaminated with such dyes, resulting in an international ban on incorporation of these dyes into food products. Shell nanocrystalline-palladium/gold core–based nanomaterials and cadmium sulfide quantum dots/polyamidoamine dendrimers were employed to amplify signals for Sudan I identification in an electro-chemiluminescent–based immunosensor (Wang et al., 2018). Dopamine powder is used to enhance the building of muscle in livestock, but as dopamine has been linked to various psychiatric and neurological diseases like schizophrenia and Parkinson's disease, many countries have prohibited its use for animal feeds. A calorimetric assay was reported by Bülbül et al. (2016) that utilized redox active nanoceria and found that dopamine can be oxidized by CeO_2 nanoparticles, and a complex of colored charge transfer was produced as a product of the oxidative reaction, which can be observed easily with the help of a spectroscope. Furthermore, for dopamine estimation, a mix of TiO_2 nanoparticles and CeO_2 were immobilized along with tyrosinase to fabricate an electrochemical biosensor with improved sensitivity and amplified signals with a detection limit of 1 nm (Njagi et al., 2010).

10.6.4 Residual Veterinary Antibiotics

For enhancement of growth and to treat animal disease, a variety of veterinary drugs like growth promoters, antiparasitic drugs and antimicrobial drugs are being used. Such drugs have the potential to accumulate in animal meat and products, leading to the presence of drug residues that further raise concerns about safety. Instant determination methods for antibiotic content residue are required at present to identify the presence of antibiotics at the point of their use (Wu et al., 2006) A gold- and silver-based SERS along with a DNA aptamer was immobilized covalently on gold nanoparticles; then a coating with a silver shell was studied for determination of kanamycin in milk. In honey, 26 kinds of sulfonamides antibiotics were tested using a lateral flow device based on a silver nanoparticle immune system. Silver nanoparticles are conjugated with a lateral flow device. The HPLC technique was used to validate the performance of the developed sensor on commercial honey (Chen et al., 2017).

10.6.5 Heavy Metals

Heavy metal contamination due to the environment leads to further food contamination. Images and text used to label food packaging materials are also a source of contamination by heavy metals (Turner, 2019). The presence of lead (Pb^{2+}), arsenic (As^{3+}) and mercury (Hg^{2+}) pose a significant concern for public health. Traditional methods include inductively coupled plasma mass spectroscopy and atomic absorption/emission spectroscopy for analysis of heavy metals. The given methods are highly sensible and offer low limits of detection, but due to their low availability and high cost, their application is limited. The use of nanomaterials in biosensors overcomes such limitations and enables field monitoring of heavy metal contamination. Anodic striping voltammetry with electrochemical sensors serves as an user-friendly, cost-effective and simple alternative for heavy metal analysis.

Nanomaterials like carbon nanotubes and graphene integrated with different detection elements like enzyme, antibodies and aptamers provide a high surface area for detection of heavy metals (Xu et al., 2018). Gold electrodes covered with a thiphenol-functionalized single-walled CNT-based sensor were fabricated for identification of Hg^{2+}, exhibiting a detection limit of 3 nM (Wei et al., 2014). A magnetic nanocomposite coated with glutathione was developed by Baghayeri M et al. (2018) for the detection of Cd^{2+} and Pb^{2+} with anodic stripping voltammetry. The developed nanocomposite facilitated metal ion reduction and pre-concentration at a modified glass carbon electrode with a reported detection limit of 0.182 and 0.172 $\mu g\ L^{-1}$.

10.6.6 Industrial Applications in the Food Sector

Along with preservation perspectives, the monitoring of food products is a new horizon for the food industry related to nanotechnology. These systems are able to respond to environmental conditions and alert the consumer about the food quality, whether it is contaminated, adulterated or spoiled. These are known as nanobiosensors and provide immediate irreversible visual qualitative information about the packaged food by diffusion of dye, change in intensity of color or color change. Such systems combine biology with nanotechnology in sensors that actually hold the potential to accurately detect contaminants with increased sensitivity and reduced response time. These nanobiosensors started in the biomedical field, but now they are not restricted to the field of medicine in detecting, monitoring and diagnosing disease, they are used in the field of the environment in detecting toxins, pollutants and heavy metals and in the recognition of foodborne pathogens (Sagadevan and Periasamy, 2014). In addition, nanobiosensors also find application in food industry in maintaining food safety and to extend the shelf life of products (Figure 10.3). This is known as smart packaging.

Nanobiosensors in food application are divided into two categories: external and internal. External nanosensors detect the impact of the external environment on products from farm to fork, while internal biosensors detect chemical and microorganism contaminants inside the package (Davor et al., 2011). Nanobiosensors are produced by putting a chemical or bacterial detector on a probe, which, in reaction with a targeted chemical or bacteria, causes a change in the nanosensor output (Kampers, 2008). Use of nanobiosensors in packaging will result in detection of spoilage along with pathogenic microorganisms in the product by changing the color of the package, while changes in chemical elements like high concentrations of oxygen are also possible (Liu et al., 2008).

Nanobiosensors play a significant role in detecting traces of adulterants, pesticides, toxins, organic compounds and even very low

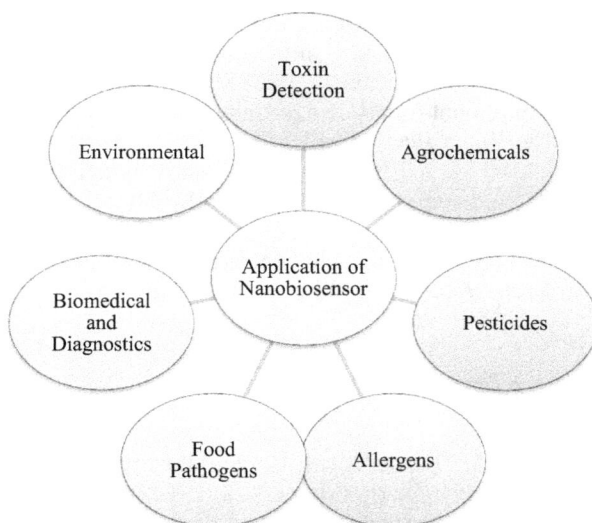

Figure 10.3 Applications of nanobiosensors.

concentrations of pathogens and harmful chemicals (Rai et al., 2012). They are also used to recognize the integrity of packages and freshness of food (Kuswandi and Moradi, 2019). These nanosensors finds application over traditional techniques like liquid/gas chromatography with mass spectroscopy due to the availability of high surface-to-volume ratios, great sensitivity in the interface, excellent optical and electric properties, lower recognition limit, quick response, specificity and selectivity (Duncan, 2011; Shivakumar et al., 2018).

10.6.6.1 Nanobiosensors in Detecting Toxins

Toxins produced by pathogenic bacteria must be analyzed in order to ensure food safety and quality, as well as to eradicate any form of human health risk. Mycotoxins are secondary metabolites produced by molds that represent a health concern to humans. Various mycotoxins, like aflatoxin, ochratoxin, patulin, fumonisins and zearalenone, are present in foodstuffs. A microfluidic sensor is a type of nanobiosensor based on microfluidics in conjunction with liposomes, which helps in detecting toxins in aqueous samples (Thakur and Ragavan, 2013). Electrochemical biosensors based on nanomaterials such as metallic nanoparticles (zinc, copper, silver, gold and platinum), carbon nanotubes (single and multiwalled) and super-paramagnetic nanoparticles are now being utilized to detect various toxins found in food products (Rai et al., 2012; Xiang et al., 2011). This will find the presence of toxins in a food product, alert the consumer about its consumption and decrease the chances of foodborne illnesses.

Speroni et al. (2010) developed a nanobiosensor-based ELISA for the identification of Ara h3/4 allergen in peanuts using protein A-coated magnetic microparticles. In order to detect aflatoxin B1, gold nanoparticles containing anti-aflatoxin antibodies were utilized. Similarly, anti-aflatoxin M1 antibodies in super-paramagnetic nanoparticles have been used to detect aflatoxin M1 in milk samples. (Radoi et al., 2008; Sharma et al., 2010; Xiulan et al., 2005; Zhang et al., 2007). Likewise, gold nanoparticles have also been used to detect brevetoxins and botulinum neurotoxin type B in processed foods using immune chromatographic and enzyme-linked immunosorbent assays (Zhang et al., 2007; Zhou et al., 2009). Generally, seafood is contaminated with a marine toxin called palytoxin, which can be detected using an electro-chemiluminescent carbon nanotube-based sensor (Wang et al., 2009; Zamolo et al., 2012).

Mycotoxins, fungal metabolites, are highly toxic in nature and cause immense toxicity even at low levels of exposure. At present, no appropriate method or technique for field identification of mycotoxins has been found, so early detection of mycotoxin contamination is quite difficult (Cheli et al., 2008). Ochratoxin A (OTA), one of the most abundant mycotoxins, is produced by *Penicillium verrucosum*, and *Aspergillus ochraceus* is a nephrotoxin with a prominent carcinogenic and immunotoxic influence on human beings. A sensor based on CeO_2 nanoparticles was fabricated by Bülbül et al. (2016) to quantify OTA. Functionalized CeO_2 nanoparticles were assembled with a biosensor with OTA-specific ssDNA aptamers ensuing better activity and dispersibility. On binding of ssDNA with its target, any difference in the redox properties on the surface of CeO_2 was determined using TMB, allowing quick visual identification of OTA. This system was found to be very receptive with a detection limit of OTA as low as 0.15 nM.

10.6.6.2 Nanobiosensors in Detecting Food Pathogens

Contamination by spoilage-causing and harmful microorganisms can occur at any time during procurement, processing, packaging, storage, transportation, retailing or even at home during meal preparation and storage. This contamination can cause a variety of foodborne illnesses that can be dangerous or even fatal. As a result, a quick and sensitive technique for identifying and detecting foodborne pathogens in food samples is required. There are a variety of ways to detect harmful microorganisms, including culture methods, immunologically based biosensors and so on. However, these methods have the disadvantages of being time consuming, labor intensive and lacking in sensitivity and specificity. There are molecular-based techniques that overcome these disadvantages, for instance, PCR and real-time PCR, but these are designed for the detection of a single species of microorganism in one assay and require highly qualified trained personnel. Nanobiosensors are being utilized to detect viruses these days since they are incredibly particular and sensitive, and they can be directly inserted into food packaging. The interaction of a biosensing device with a nanomaterial and a pathogen is used to recognize the pathogen, which is then monitored using a signal transduction process.

Nanosensors are devices that monitor the quality of food processing and food safety. When biology and nanoscale technology are joined in the production of sensors, the result is a nanobiosensor that can detect pollutants and pathogens. In this case, nanoparticles bind to the food pathogen and are detected by sensors that use magnetic or infrared light. Various nanoparticles can be attached to a single nanosensor, allowing for rapid and reliable detection of multiple germs or diseases. Time is compressed from days to hours, minutes and even seconds using this technology (Choudhury and Goswami, 2012). The target recognition groups for nanostructures as biosensors are biological substances such as sugar and proteins (Charych et al., 1996). These are primarily intended for use in identifying food contaminants and pathogens and tracking food products.

Identification of the genetic material or whole bacterial cell help in detecting foodborne pathogens in food material. Conventional methods already exist to identify pathogens, and they are reliable but complicated at the same time (Kumar et al., 2018). Therefore, to overcome this, nanobiosensors find application in detecting pathogens immediately. They are being attached to the packaging of food. A nanobioluminescent spray made up of several magnetic nanoparticles reacts with pathogens in food to produce a visible glow that may be identified easily (Arshak et al., 2007). *Listeria monocytogenes* DNA was isolated from milk using a magnetic iron oxide-based nanobioluminescent spray (Yang et al., 2010). One of the most effective ways for detecting distinct foodborne pathogenic bacteria is Raman spectroscopy combined with silver nanosensors (Fang et al., 2017). Other nanomaterials, such as nanocolloids, nanorods, plasmonic gold, graphene oxide, magnetic beads and carbon nanotubes, are used to detect foodborne infections in addition to silver (Baranwal et al., 2016; Holzinger et al., 2014; Kahraman et al., 2008; Zuo et al., 2013). For the identification of *Escherichia coli, Listeria monocytogenes* and *Salmonella* species, an array-based immunosorbent test combined with liposomal nanovesicles has recently found favor (Chen et al., 2011; Ghasemi-Varnamkhasti et al., 2011; Sondi and Salopek-Sondi, 2004). Furthermore, silicon-based nanobiosensors paired with proteins have been used in a variety of liquid food systems (Wang et al., 2011). Pathogens can be discovered in diverse food samples using DNA in industry. Silver (Au) nanoparticles have been coupled with a DNA-sensing device to improve the sensitivity and selectivity for the target DNA (Zeng et al., 2007).

10.6.6.3 Nanobiosensors in Detecting Chemicals and Pesticides

Pesticide residues are one of the most serious issues in the food sector. Because of their great insecticidal activity and low persistence, carbamate and organophosphate chemicals are the most extensively used pesticides (Mora et al., 1996). These herbicides block acetylcholinesterase, causing nerve transmission to be interrupted, which is a major public health risk. As a result, highly sensitive and specific ways to detect the presence of pesticides in a short period of time are required, and nanobiosensors precisely meet this requirement. Colorimetric and fluorometric nanosensors conjugated with gold nanoparticles are being utilized to detect organophosphorus and carbamate insecticides (Liu et al., 2012). Potentiometer sensors made of silica nanocomposites and multi-walled carbon nanotubes are used to detect hazardous cadmium ions (Bagheri et al., 2013). Paraoxon sensing has been detected using core-shell quantum dots constructed of cadmium selenide and zinc sulphate (Ji et al., 2005). It has been claimed that volumetric and fluorescent nanosensors have been developed for the detection of parathion and melamine herbicides (Varshney et al., 2007; Wang and Irudayaraj, 2008). Similarly, ionic-liquid nanocomposite modified with multiwalled carbon nanotubes has been used to detect dyes such as Sudan-I, sunset yellow and tartrazine. For the detection of bisphenol A and Sudan I, another nanocomposite containing zinc oxide nanoparticles and carbon nanotubes was used (Sanchez-Acevedo et al., 2009; Najafi et al., 2014).

Pesticides include a range of chemicals such as herbicides, plant growth regulators, fungicides and insecticides and are used for crop protection and to increase production of food. But such pesticides leave residues in water, plants and soil on application and pose a health hazard (Aktar et al., 2009) A non-enzymatic approach was used to detect pesticides like fenthion and chlorpyrifos with the help of CuO nanostructures developed *in situ* on indium tin oxide (ITO) (Tunesi et al., 2018). The detection principle was based on the fact that on the surface of ITO, a coating of pralidome chloride on CuO nanostructures functionalized by pimelic acid was inhibited. Improved signal due to higher loading and surface area was attained with the help of an in situ nanostructure assembly. For detection of organophosphates, a printable, disposable graphene based biosensor was fabricated using patterned graphene where inkjet maskless lithography was used for the modification of electrodes with subsequent laser annealing, PtNP electrodeposition and phosphotriesterase ink casting. Increased loading, higher surface area and conductivity enhanced sensitive and instant identification of paraoxon in samples of water and soil (Hondred et al., 2018).

10.7 Advantages of Nanobiosensors

A tremendous technology push have been generated due to several benefits of nanobiosensors. They are very rapid and ultrasensitive techniques for detecting foodborne microorganisms and toxins. They are a cost-effective, time-efficient and non-destructive method compared to traditional analytical methods. The sensors are bendable, lightweight, rollable, thin, compact, more sensitive, selective and reversible, with optical and electrical qualities that are comparable or superior to those of conventional sensors (Kiss, 2020). The following are some of the other advantages of nanobiosensors:

- They are target specific, which is a key factor in early detection.
- Rapid and high detection throughput.
- The analytical process is simple, fast, cost effective and user friendly.
- They are ultrasensitive, with good repeatability, and are portable and stable.
- Reduced material requirements, and they are recyclable.
- They are non-destructive, refined and reliable.

10.8 Future Perspectives and Conclusion

Nanobiosensors are one of the most studied fields of nanobiotechnology, since they are used to detect pesticide, adulterant, poisons, organic compounds and even very low concentrations of pathogens and dangerous substances. They can also be used in the packaging, processing, preservation, and safety of food. They provide a secure method of delivering bioactive chemicals. Nanobiosensors have several applications in the food industry with numerous benefits; yet, due to their cytotoxic effects, they pose a major hazard to human health and the environment (Shafiq et al., 2020). Because of their small size and subcellular interaction with other components, nanomaterials pose a health risk. Nanoparticles, for example, permeate the skin and create a variety of health concerns in both animals and humans. Similarly, nanomaterials affect plant growth by altering protein and gene levels (Rizwan et al., 2017). As a result, it's critical to understand the mechanisms through which distinct nanomaterials interact with human health and the environment. We need to figure out how nanoparticles get into the human body from food. Nanomaterials are consumed, either intentionally or accidentally, through the intraoral, dermal and respiratory pathways, posing a major health danger to people (Momin et al., 2013; Raynes et al., 2014). To summarize, nanotechnology has started a revolution in the food and packaging industries due to advanced nanomaterials and nanodevices, despite the fact that it is still in its early stages. In-depth study of nanobiosensors is still required by keeping of the safety of consumers and proper toxicological study trials in view.

References

Agrawal, S., & Prajapati, R. (2012). Nanosensors and their pharmaceutical applications: A review. *International Journal of Pharmaceutical Science and Technology. 4*, 1528–1535.

Aktar, M.W., Sengupta, D., & Chowdhury, A. (2009). Impact of pesticides use in agriculture: Their benefits and hazards. *Interdisciplinary Toxicology. 2*, 1.

Andrei, V., Sharpe, E., Vasilescu, A., & Andreescu, S. (2016). A single use electrochemical sensor based on biomimetic nanoceria for the detection of wine antioxidants. *Talanta, 156*, 112–118

Arshak, K., Adley, C., Moore, E., Cunni_e, C., Campion, M., & Harris, J. (2007). Characterisation of polymernanocomposite sensors for quantification of bacterial cultures. *Sensors and Actuators B: Chem. 126*, 226–231.

Baghayeri, M., Amiri, A., Maleki, B., Alizadeh, Z., & Reiser, O. (2018). A simple approach for simultaneous detection of cadmium (II) and lead (II) based on glutathione coated magnetic nanoparticles as a highly selective electrochemical probe. *Sensors and Actuators B: Chemical. 273*, 1442–1450.

Bagheri, H., Afkhami, A., Shirzadmehr, A., Khoshsafar, H., Khoshsafar, H., & Ghaedi, H. (2013). Novel potentiometric sensor for the determination of Cd^{2+} based on a new nano-composite. *International Journal of Environment and Analytical Chemistry. 93*, 578–591.

Bahadır, E.B., & Sezgintürk, M.K. (2016). Poly (amidoamine)(PAMAM): An emerging material for electrochemical bio (sensing) applications. *Talanta. 148*, 427–438.

Baranwal, A., Mahato, K., Srivastava, A., Maurya, P.K., & Chandra, P. (2016). Phytofabricated metallic nanoparticles and their clinical applications. *RSC Advances. 6*, 105996–106010.

Bhushan, B. (2007). Nanotribology and nanomechanics of MEMS/NEMS and BioMEMS/BioNEMS materials and devices. *Microelectronic Engineering. 84*, 387–412.

Bülbül, G., Hayat, A., & Andreescu, S. (2016). ssDNA-functionalized nanoceria: A redox-active aptaswitch for biomolecular recognition. *Advanced Healthcare Materials. 5*, 822–828.

Chang, A.C., Gillespie, J.B., & Tabacco, M.B. (2001). Enhanced detection of live bacteria using a dendrimer thin film in an optical biosensor. *Analytical Chemistry. 73*, 467–470.

Charych, D., Cheng, Q., Reichert, A., Kuziemko, G., Stroh, N., Nagy, J., & Spevak, W. (1996). A 'litmus test' for molecular recognition using artificial membranes. *Chemical & Biology. 3*, 113–120.

Cheli, F., Pinoti, L., Campagnoli, A., Fusi, E., Rebuci, R., & Baldi, A. (2008). Mycotoxin analysis, mycotoxin producing fungi assays and mycotoxin toxicity bioassays in food mycotoxin monitoring and surveillance. *Italian Journal of Food Science, 20*.

Chen, S., Ma, L., Yuan, R., Chai, Y., Xiang, Y., & Wang, C. (2011). Electrochemical sensor based on Prussian blue nanorods and gold nanochains for the determination of H2O2. *European Food and Research Technology. 232*, 87–95.

Chen, Y., Liu, L., Xu, L., Song, S., Kuang, H., Cui, G., & Xu, C. (2017). Gold immunochromatographic sensor for the rapid detection of twenty-six sulfonamides in foods. *Nano Research. 10*, 2833–2844.

Choudhury, S.R., & Goswami, A. (2012) Supramolecular reactive sulphur nanoparticles: A novel and efficient antimicrobial agent. *Journal of Applied Microbiology. 114*, 1–10.

Davis, D., Guo, X., Musavi, L., Lin, C.S., Chen, S.H., & Wu, V.C.H. (2013). Gold nanoparticle-modified carbon electrode biosensor for the detection of *Listeria monocytogenes. Industrial Biotechnology. 9*, 31–36.

Davor, I., Goran, M., & Dražan, K. (2011). Review of potential use, benefits and risks of nanosensors nanotechnologies in food. *Strojarstvo. 53*, 127–136.

Duncan, T.V. (2011). Applications of nanotechnology in food packaging and food safety: Barrier materials, antimicrobials and sensors. *Journal of Colloid and Interface Sciences. 363*, 1–24.

El-Nour, K.M.A., Salam, E.T.A., Soliman, H.M., & Orabi, A.S. (2017). Gold nanoparticles as a direct and rapid sensor for sensitive analytical detection of biogenic amines. *Nanoscale Research Letters. 12*, 1–11.

Fang, Z., Zhao, Y., Warner, R.D., & Johnson, S.K. (2017). Active and intelligent packaging in meat industry. *Trends in Food Science and Technology. 61*, 60–71.

Faridbod, F., Norouzi, P., Dinarvand, R., & Ganjali, M.R. (2008). Developments in the field of conducting and non-conducting polymer based potentiometric membrane sensors for ions over the past decade. *Sensors. 8*, 2331–2412.

Foster, L.E. (2006). *Nanotechnology: Science, Innovation and Opportunity*; Prentice Hall: Upper Saddle River, NJ, p. 283.

Geise, R.J., Adams, J.M., Barone, N.J., & Yacynych, A.M. (1991). Electropolymerized films to prevent interferences and electrode fouling in biosensors. *Biosensors and Bioelectronics. 6*, 151–160.

Ghasemi-Varnamkhasti, M., Mohtasebi, S., Rodriguez-Mendez, M., Siadat, M., Ahmadi, H., & Razavi, S. (2011). Electronic and bioelectronic tongues, two promising analytical tools for the quality evaluation of nonalcoholic beer. *Trends in Food Science and Technology. 22*, 245–248.

Guo, S., & Wang, E. (2007). Synthesis and electrochemical applications of gold nanoparticles. *Analytica Chimica Acta. 598*, 181–192.

Hochella Jr, M. F. (2002). Nanoscience and technology: The next revolution in the Earth sciences. *Earth and Planetary Science Letters.* *203*, 593–605.

Holzinger, M., Le Go, A., & Cosnier, S. (2014) Nanomaterials for biosensing applications: A review. *Frontier Chemistry.* *2*, 63.

Hondred, J.A., Breger, J.C., Alves, N.J., Trammell, S.A., Walper, S.A., Medintz, I.L., & Claussen, J.C. (2018). Printed graphene electrochemical biosensors fabricated by inkjet maskless lithography for rapid and sensitive detection of organophosphates. *ACS Applied Materials & Interfaces.* *10*, 11125–11134.

Ji, X., Zheng, J., Xu, J., Rastogi, V.K., Cheng, T.C., DeFrank, J.J., & Leblanc, R.M. (2005). (CdSe) ZnS quantum dots and organophosphorus hydrolase bioconjugate as biosensors for detection of paraoxon. *The Journal of Physical Chemistry B.* *109*, 3793–3799.

Jianrong, C., Yuqing, M., Nongyue, H., Xiaohua, W., & Sijiao, L. (2004). Nanotechnology and biosensors. *Biotechnology advances,* *22*(7), 505–518.

Kahraman, M., Yazıcı, M.M., Sahin, F., & Çulha, M. (2008). Convective assembly of bacteria for surface-enhanced Raman scattering. *Langmuir.* *24*, 894–901.

Kampers, F. (2008). Nanotechnologies for food. In *Proceedings of the Nanotech Northern Europe Conference*; Copenhagen, Denmark, September 23–25.

Kasry, A., Ardakani, A.A., Tulevski, G.S., Menges, B., Copel, M., & Vyklicky, L. (2012). Highly efficient fluorescence quenching with graphene. *The Journal of Physical Chemistry C.* *116*, 2858–2862.

Khan, R., Rehman, A., Hayat, A., & Andreescu, S. (2019). Magnetic particles-based analytical platforms for food safety monitoring. *Magnetochemistry.* *5*, 63.

Kiss, E. (2020). Nanotechnology in food systems: A review. *Acta Alimentaria.* *49*, 460–474.

Kissinger, P.T. (2005). Biosensors—a perspective. *Biosensors and Bioelectronics.* *20*, 2512–2516.

Kumar, S., Shukla, A., Baul, P.P., Mitra, A., & Halder, D. (2018). Biodegradable hybrid nanocomposites of chitosan/gelatin and silver nanoparticles for active food packaging applications. *Food Packaging and Shelf Life.* *16*, 178–184.

Kuswandi, B., & Moradi, M. (2019). Improvement of food packaging based on functional nanomaterial. In *Nanotechnology: Applications in Energy, Drug and Food*; Springer: Cham, Switzerland, pp. 309–344.

Li, H., Kang, Z., Liu, Y., & Lee, S.T. (2012). Carbon nanodots: Synthesis, properties and applications. *Journal of Materials Chemistry.* *22*, 24230–24253.

Liu, C., Jia, Q., Yang, C., Qiao, R., Jing, L., Wang, L., & Gao, M. (2011). Lateral flow immunochromatographic assay for sensitive pesticide detection by using Fe3O4 nanoparticle aggregates as color reagents. *Analytical Chemistry.* *83*, 6778–6784.

Liu, D., Chen, W., Wei, J., Li, X., Wang, Z., & Jiang, X. (2012). A highly sensitive, dual-readout assay based on gold nanoparticles for organophosphorus and carbamate pesticides. *Analytical Chemistry. 84*, 4185–4191.

Liu, S., Yuan, L., Yue, X., Zheng, Z., & Zhiyong, T. (2008). Recent advances in nanosensors for organophosphate pesticide detection. *Advanced Powder Technology. 19*, 419–441.

Malhotra, B.D., Chaubey, A., & Singh, S.P. (2006). Prospects of conducting polymers in biosensors. *Analytica Chimica Acta. 578*, 59–74.

Malik, P., Katyal, V., Malik, V., Asatkar, A., Inwati, G., & Mukherjee, T.K. (2013). Nanobiosensors: Concepts and variations. *International Scholarly Research Notices, 2013*.

Mathew, F.P., & Alocilja, E.C. (2005). Porous silicon-based biosensor for pathogen detection. *Biosensors and Bioelectronics. 20*, 1656–1661.

Momin, J.K., Jayakumar, C., & Prajapati, J.B. (2013). Potential of nanotechnology in functional foods. *Emirates Journal of Food and Agriculture. 25*.

Mora, A., Comejo, J., Revilla, E., & Hermosin, M.C. (1996). Persistence and degradation of carofuran in Spanish soil suspensions. *Chemosphere. 32*, 1585–1598.

Muhammad-Tahir, Z., & Alocilja, E.C. (2003). Fabrication of a disposable biosensor for *Escherichia coli* O157: H7 detection. *IEEE Sensors Journal. 3*, 345–351.

Muhammad-Tahir, Z., & Alocilja, E.C. (2004). A disposable biosensor for pathogen detection in fresh produce samples. *Biosystems Engineering. 88*, 145–151.

Mustafa, Fatima, and Silvana Andreescu. (2020a). Based enzyme biosensor for one-step detection of hypoxanthine in fresh and degraded fish. *ACS Sensors 5.12*: 4092–4100.

Mustafa, F., & Andreescu, S. (2020b). Nanotechnology-based approaches for food sensing and packaging applications. *RSC Advances, 10*(33), 19309–19336.

Najafi, M., Khalilzadeh, M.A., & Karimi-Maleh, H. (2014). A new strategy for determination of bisphenol a in the presence of Sudan I using a ZnO/CNTs/ionic liquid paste electrode in food samples. *Food Chemistry. 158*, 125–131.

Njagi, J., Chernov, M.M., Leiter, J.C., & Andreescu, S. (2010). Amperometric detection of dopamine *in vivo* with an enzyme based carbon fiber microbiosensor. *Analytical Chemistry. 82*(3), 989–996.

Pak, S.C., Penrose, W., & Hesketh, P.J. (2001). An ultrathin platinum film sensor to measure biomolecular binding. *Biosensors and Bioelectronics. 16*, 371–379.

Park, J.H., Gu, L., Von Maltzahn, G., Ruoslahti, E., Bhatia, S.N., & Sailor, M.J. (2009). Biodegradable luminescent porous silicon nanoparticles for *in vivo* applications. *Nature Materials. 8*, 331–336.

Phillips, E., Penate-Medina, O., Zanzonico, P.B., Carvajal, R.D., Mohan, P., Ye, Y., & Bradbury, M.S. (2014). Clinical translation of an ultrasmall inorganic optical-PET imaging nanoparticle probe. *Science Translational Medicine. 6*, 26049–260149.

Pumera, M. (2011). Graphene in biosensing. *Materials Today, 14*(7–8), 308–315.

Radoi, A., Targa, M., Prieto-Simon, B., & Marty, J.-L. (2008). Enzyme-Linked immunosorbent assay (ELISA) based on superparamagnetic nanoparticles for aflatoxin M1 detection. *Talanta, 77*, 138–143.

Rai, M., Gade, A., Gaikwad, S., Marcato, P.D., & Durán, N. (2012). Biomedical applications of nanobiosensors: The state-of-the-art. *Journal of the Brazilian Chemical Society. 23*, 14–24.

Rao, C.E.E., Sood, A.E., Subrahmanyam, K.E., & Govindaraj, A. (2009). Graphene: The new two-dimensional nanomaterial. *Angewandte Chemie International Edition. 48*, 7752–7777.

Raynes, J.K., Carver, J.A., Gras, S.L., & Gerrard, J.A. (2014). Protein nanostructures in food-Should we be worried. *Trends in Food Science and Technology. 37*, 42–50.

Rizwan, M., Ali, S., Qayyum, M.F., Ok, Y.S., Adrees, M., Ibrahim, M., Zia-ur-Rehman, M., Farid, M., & Abbas, F. (2017). Effect of metal and metal oxide nanoparticles on growth and physiology of globally important food crops: A critical review. *Journal of Hazardous Materials. 322*, 2–16.

Sagadevan, S., & Periasamy, M. (2014). Recent trends in nanobiosensors and their applications—a review. *Review on Advanced Materials Science. 36*, 62–69.

Sánchez-Acevedo, Z.C., Riu, J., & Rius, F.X. (2009). Fast picomolar selective detection of bisphenol a in water using a carbon nanotube field e_ect transistor functionalized with estrogen receptor. *Biosensors and Bioelectronics. 24*, 2842–2846.

Shafiq, M., Anjum, S., Hano, C., Anjum, I., & Abbasi, B.H. (2020). An overview of the applications of nanomaterials and nanodevices in the food industry. *Foods. 9*, 148–175.

Sharma, A., Matharu, Z., Sumana, G., Solanki, P.R., Kim, C., & Malhotra, B. (2010). Antibody immobilized cysteamine functionalized-gold nanoparticles for aflatoxin detection. *Thin Solid Films. 519*, 1213–1218.

Sheikhzadeh, E., Chamsaz, M., Turner, A.P.F., Jager, E.W.H., & Beni, V. (2016). Label-free impedimetric biosensor for *Salmonella Typhimurium* detection based on poly [pyrrole-co-3-carboxyl-pyrrole] copolymer supported aptamer. *Biosensors and Bioelectronics. 80*, 194–200.

Shiddiky, M.J., Rahman, M.A., Cheol, C.S., & Shim, Y.B. (2008). Fabrication of disposable sensors for biomolecule detection using hydrazine electrocatalyst. *Analytical Biochemistry. 379*, 170–175.

Shiddiky, M.J., Rahman, M.A., & Shim, Y.B. (2007). Hydrazine-catalyzed ultrasensitive detection of DNA and proteins. *Analytical Chemistry. 79*, 6886–6890.

Shivakumar, N., Madhusudan, P., & Daniel, S.K. (2018). Nanomaterials for smart food packaging. In *Handbook of Nanomaterials for Industrial Applications*; Elsevier: Amsterdam, The Netherlands, pp. 260–270.

Sondi, I., & Salopek-Sondi, B. (2004). Silver nanoparticles as antimicrobial agent: A case study on E. coli as a model for Gram-negative bacteria. *Journal of Colloid Interface and Sciences. 275*, 177–182.

Song, X., Wang, D., & Kim, M. (2021). Development of an immuno-electrochemical glass carbon electrode sensor based on graphene oxide/gold nanocomposite and antibody for the detection of patulin. *Food Chemistry, 342,* 128257.

Sperling, R.A., Gil, P.R., Zhang, F., Zanella, M., & Parak, W.J. (2008). Biological applications of gold nanoparticles. *Chemical Society Reviews. 37,* 1896–1908.

Speroni, F., Elviri, L., Careri, M., & Mangia, A. (2010). Magnetic particles functionalized with PAMAM-dendrimers and antibodies: A new system for an ELISA method able to detect Ara h3/4 peanut allergen in foods. *Analytical and Bioanalytical Chemistry. 397,* 3035–3042.

Thakur, M., & Ragavan, K. (2013). Biosensors in food processing. *Journal of Food Science and Technology. 50,* 625–641.

Tiwari, I., Singh, M., Pandey, C.M., & Sumana, G. (2015). Electrochemical genosensor based on graphene oxide modified iron oxide–chitosan hybrid nanocomposite for pathogen detection. *Sensors and Actuators B: Chemical. 206,* 276–283.

Tully, E., Higson, S.P., & O'Kennedy, R. (2008). The development of a 'labeless' immunosensor for the detection of *Listeria monocytogenes* cell surface protein, Internalin B. *Biosensors and Bioelectronics. 23,* 906–912.

Tunesi, M.M., Kalwar, N., Abbas, M.W., Karakus, S., Soomro, R.A., Kilislioglu, A., & Hallam, K.R. (2018). Functionalised CuO nanostructures for the detection of organophosphorus pesticides: A non-enzymatic inhibition approach coupled with nano-scale electrode engineering to improve electrode sensitivity. *Sensors and Actuators B: Chemical. 260,* 480–489.

Turner, A. (2019). Heavy metals in the glass and enamels of consumer container bottles. *Environmental Science & Technology. 53,* 8398–8404.

Valko, M., Rhodes, C. J. B., Moncol, J., Izakovic, M. M., & Mazur, M. (2006). Free radicals, metals and antioxidants in oxidative stress-induced cancer. *Chemico-biological interactions, 160*(1), 1–40.

Van Gerwen, P., Laureyn, W., Laureys, W., Huyberechts, G., De Beeck, M.O., Baert, K., & Mertens, R. (1998). Nanoscaled interdigitated electrode arrays for biochemical sensors. *Sensors and Actuators B: Chemical. 49,* 73–80.

Varshney, M., Li, Y., Srinivasan, B., & Tung, S. (2007). A label-free, microfluidics and interdigitated array microelectrode-based impedance biosensor in combination with nanoparticles immunoseparation for detection of Escherichia coli O157: H7 in food samples. *Sensors and Actuators B: Chemical. 128,* 99–107.

Velusamy, V., Arshak, K., Korostynska, O., Oliwa, K., & Adley, C. (2010). An overview of foodborne pathogen detection: In the perspective of biosensors. *Biotechnology Advances. 28,* 232–254.

Wang, B., Chen, Y., Wu, Y., Weng, B., Liu, Y., Lu, Z., & Yu, C. (2016). Aptamer induced assembly of fluorescent nitrogen-doped carbon dots on gold nanoparticles for sensitive detection of AFB1. *Biosensors and Bioelectronics. 78,* 23–30.

Wang, C., Hu, L., Zhao, K., Deng, A., & Li, J. (2018). Multiple signal amplification electrochemiluminescent immunoassay for Sudan I using gold
nanorods functionalized graphene oxide and palladium/aurum core-
shell nanocrystallines as labels. *Electrochimica Acta. 278*, 352–362.

Wang, C., & Irudayaraj, J. (2008). Gold nanorod probes for the detection of
multiple pathogens. *Small. 4*, 2204–2208.

Wang, D.B., Tian, B., Zhang, Z.P., Wang, X.Y., Fleming, J., Bi, L.J., & Zhang,
X.E. (2015). Detection of Bacillus anthracis spores by super-paramagnetic lateral-flow immunoassays based on "road closure". *Biosensors
and Bioelectronics. 67*, 608–614.

Wang, L., Chen, W., Xu, D., Shim, B.S., Zhu, Y., Sun, F., Liu, L., Peng, C., Jin,
Z., & Xu, C. (2009). Simple, rapid, sensitive, and versatile SWNT-paper
sensor for environmental toxin detection competitive with ELISA.
Nano Letters. 9, 4147–4152.

Wang, L., Luo, J., Shan, S., Crew, E., Yin, J., Zhong, C.J., Wallek, B., &
Wong, S.S. (2011). Bacterial inactivation using silver-coated magnetic
nanoparticles as functional antimicrobial agents. *Analytical Chemistry.
83*, 8688–8695.

Wang, N., Pandit, S., Ye, L., Edwards, M., Mokkapati, V. R. S. S., Murugesan,
M., ... & Liu, J. (2017). Efficient surface modification of carbon nanotubes for fabricating high performance CNT based hybrid nanostructures. *Carbon, 111*, 402-410.

Wang, R., Xu, Y., Zhang, T., & Jiang, Y. (2015). Rapid and sensitive detection
of Salmonella typhimurium using aptamer-conjugated carbon dots as
fluorescence probe. *Analytica Chimica Acta. 7*, 1701–1706.

Wang, Y., Wang, T., Da, P., Xu, M., Wu, H., & Zheng, G. (2013). Silicon
nanowires for biosensing, energy storage, and conversion. *Advanced
Materials. 25*, 5177–5195.

Weber, J.E., Pillai, S., Rama, M.K., Kumar, A., & Singh, S.R. (2011).
Electrochemical impedance-based DNA sensor using a modified sin-
gle walled carbon nanotube electrode. *Material Science Engineering
C. 31*, 821–825.

Wei, J., Yang, D., Chen, H., Gao, Y., & Li, H. (2014). Stripping voltammetric
determination of mercury (II) based on SWCNT-PhSH modified gold
electrode. *Sensors and Actuators B: Chemical. 190*, 968–974.

Wu, D., Du, D., & Lin, Y. (2016). Recent progress on nanomaterial-based
biosensors for veterinary drug residues in animal-derived food. *TrAC
Trends in Analytical Chemistry. 83*, 95–101.

Xia, S., Yu, Z., Liu, D., Xu, C., & Lai, W. (2016). Developing a novel immunochromatographic test strip with gold magnetic bifunctional nanobeads (GMBN) for efficient detection of *Salmonella choleraesuis* in
milk. *Food Control. 59*, 507–512.

Xiang, L., Zhao, C., & Wang, J. (2011). Nanomaterials-Based electrochemical sensors and biosensors for pesticide detection. *Sensor Letters. 9*,
1184–1189.

Xiulan, S., Xiaolian, Z., Jian, T., Zhou, J., & Chu, F. (2005). Preparation of
gold-labeled antibody probe and its use in immunochromatography

assay for detection of aflatoxin B1. *International Journal of Food Microbiology. 99*, 185–194.

Xu, J., Cao, Z., Zhang, Y., Yuan, Z., Lou, Z., Xu, X., & Wang, X. (2018). A review of functionalized carbon nanotubes and graphene for heavy metal adsorption from water: Preparation, application, and mechanism. *Chemosphere. 195*, 351–364.

Xu, X., Ray, R., Gu, Y., Ploehn, H. J., Gearheart, L., Raker, K., & Scrivens, W. A. (2004). Electrophoretic analysis and purification of fluorescent single-walled carbon nanotube fragments. *Journal of the American Chemical Society. 126*(40), 12736–12737.

Yang, D., Zhu, L., & Jiang, X. (2010). Electrochemical reaction mechanism and determination of Sudan I at a multi wall carbon nanotubes modified glassy carbon electrode. *Journal of Electroanalytical Chemistry. 640*, 17–22.

Zamolo, V.A., Valenti, G., Venturelli, E., Chaloin, O., Marcaccio, M., Boscolo, S., Castagnola, V., Sosa, S., Berti, F., & Fontanive, G. (2012). Highly sensitive electro chemiluminescent nano biosensor for the detection of palytoxin. *ACS Nano, 6*, 7989–7997.

Zeng, X., Kong, F., Halliday, C., Chen, S., Lau, A., Playford, G., & Sorrell, C. (2007). Reverse line blot hybridization assay for identification of medically important fungi from culture and clinical specimens. *Journal of Clinical Microbiology. 45*, 2872–2880.

Zhang, D., & Alocilja, E. C. (2008). Characterization of nanoporous silicon-based DNA biosensor for the detection of *Salmonella Enteritidis. IEEE Sensors Journal. 8*, 775–780.

Zhang, H., Ma, X., Liu, Y., Duan, N., Wu, S., Wang, Z., & Xu, B. (2015) Gold nanoparticles enhanced SERS aptasensor for the simultaneous detection of *Salmonella typhimurium* and *Staphylococcus aureus. Biosensors and Bioelectronics. 15*, 872–877.

Zhang, W., Tang, H., Geng, P., Wang, Q., Jin, L., & Wu, Z. (2007). Amperometric method for rapid detection of Escherichia coli by flow injection analysis using a bismuth nano-film modified glassy carbon electrode. *Electrochemistry Communication. 9*, 833–838.

Zhang, Y. (2007) Electrochemical DNA biosensors based on gold nanoparticles/cysteamine/poly (glutamic acid) modified electrode. *American Journal of Biomedical and Research. 1*, 115–125.

Zhou, Y., Pan, F. G., Li, Y. S., Zhang, Y. Y., Zhang, J. H., Lu, S. Y., Ren, H. L., & Liu, Z. S. (2009). Colloidal gold probe-based immunochromatographic assay for the rapid detection of brevetoxins in fishery product samples. *Biosensors and Bioelectronics. 24*, 2744–2747.

Zuo, P., Li, X., Dominguez, D.C., & Ye, B.C. (2013). A PDMS/paper/glass hybrid microfluidic biochip integrated with aptamer-functionalized graphene oxide nano-biosensors for one-step multiplexed pathogen detection. *Lab on a Chip. 13*, 3921–3928.

Index

Note: Page numbers in *italics* indicate figures and those in **bold** indicate tables.

273

For Product Safety Concerns and Information please contact our EU
representative GPSR@taylorandfrancis.com
Taylor & Francis Verlag GmbH, Kaufingerstraße 24, 80331 München, Germany

www.ingramcontent.com/pod-product-compliance
Lightning Source LLC
Chambersburg PA
CBHW060341220326
41598CB00023B/2772